CRTD-Vol.76

I0041872

Technical Peer Review Report
Report of the Review Panel

Spent Nuclear Fuel Canister Welding Concept

ASME International

Institute for Regulatory Science

DISCLAIMER

This report was prepared through the collaborative efforts of the American Society of Mechanical Engineers (ASME) Center for Research and Technology Development and the Institute for Regulatory Science (referred to thereafter with the collaborators as the Society) for the Office of Science and Technology Development of the U.S. Department of Energy (referred to hereafter as the Sponsor).

Neither the Society nor the Sponsor, or others involved in the preparation or review of this report nor any of their respective employees, members, or persons acting on their behalf, make any warranty, expressed or implied, or assume any legal liability or responsibility for the accuracy, completeness, or usefulness of any information, apparatus, product, or process disclosed, or represent that its uses would not infringe privately owned rights.

Information contained in this work has been obtained by the American Society of Mechanical Engineers from sources believed to be reliable. However, neither ASME nor its authors or editors guarantee the accuracy or completeness of any information published in this work. Neither ASME nor its authors and editors shall be responsible for any errors, omissions, or damages arising out of the use of this information. The work is published with the understanding that ASME and its authors and editors are supplying information but are not attempting to render engineering or other professional services. If such engineering or professional services are required, the assistance of an appropriate professional should be sought.

Statement from By-Laws: The Society shall not be responsible for statements or opinions advanced in papers ... or printed in its publications. (7.1.3)

For authorization to photocopy material for internal or personal use under circumstances not falling within the fair use provisions of the Copyright Act, contact the Copyright Clearance Center (CCC), 222 Rosewood Drive, Danvers, MA 01923, Tel: 978-750-8400, www.copyright.com. Requests for special permission or bulk reproduction should be addressed to the ASME Technical Publishing Department.

TABLE OF CONTENTS PAGE NO.

Preface

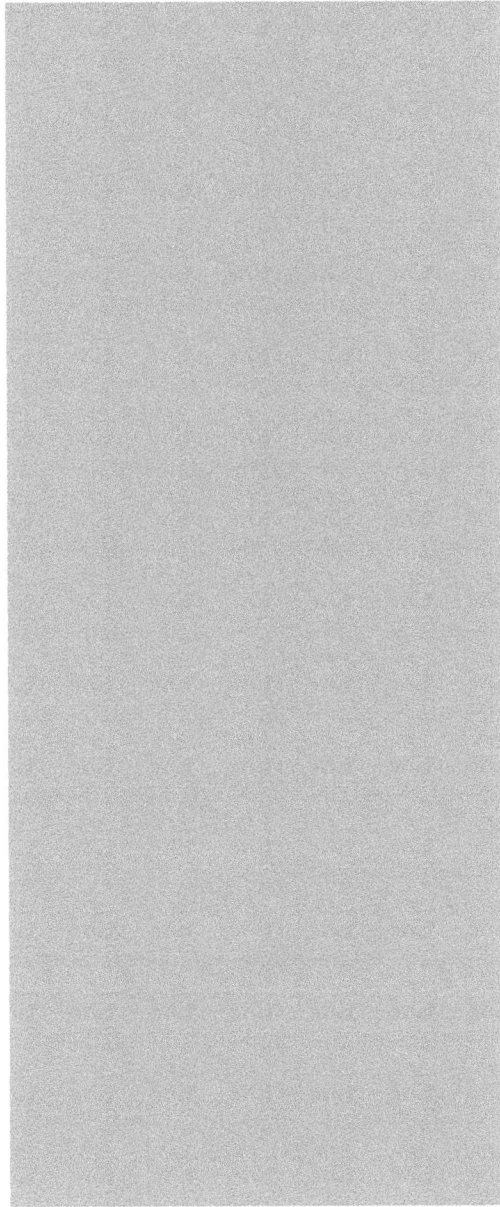

This report contains the results of a peer review performed jointly by the American Society of Mechanical Engineers (ASME) and the Institute for Regulatory Science (RSI). Based on a request from the Idaho National Engineering and Environmental Laboratory of the U.S. Department of Energy (DOE), a Review Panel (RP) was established to peer review "Spent Nuclear Fuel Canister Welding Concept". Consistent with the ASME/RSI process, the following RP was appointed by the Peer Review Committee for Energy and the Environment (PRCEE) of ASME:

George Cook
Sindo Kou
Donald Thompson
Adam Jacoff
Richard Rosenberg, Chair
Wade Troxell

During the period covered by this report, the ASME PRCEE overseeing the peer review consisted of the following individuals:

Charles O. Velzy, Member of Executive Panel (EP) and Chair
Ernest L. Daman, Member of EP
Nathan H. Hurt, Member of EP
A. Alan Moghissi, Member of EP; Principal Investigator of the Peer Review Program
Gary A. Benda
Erich W. Bretthauer
Irwin Feller
Robert A. Fjeld
William T. Gregory, III
Peter B. Lederman
Jeffrey A. Marqusee
Lawrence C. Mohr, Jr.
Goetz K. Oertel
Glen W. Suter, II
Cheryl A. Trottier

The supporting staff included the following individuals:

Michael Tinkelman: Director of Research at the Center for Research and Technology Development of ASME in Washington, DC; and Administrative Manager of the ASME PRCEE.

Betty R. Love: Executive Vice President, RSI, Columbia, MD; and Administrative Manager of the Peer Review Program.

M.C. Kirkland: Vice President for Southeastern Region, RSI.

Sorin R. Straja: Vice President for Science and Technology, RSI; and Technical Secretary.

Sharon D. Jones: Director of Training Programs, RSI; and Manager of Review Panel Operations.

The biographical summaries of the members of the RP, the PRCEE, and the technical staff are located at the end of this report.

The Review Criteria were provided by the Project Team—consisting of principal investigators, project managers, and others involved in the project. These criteria were slightly revised by the Technical Secretary and approved by the Project Team. The RP received documents describing various aspects of the project. The summary of the project included in this report was prepared by the Technical Secretary using the same documents that had been provided to the Review Panel. The Project Team received the project summary for review and approval. In addition, the staff of RSI undertook the task of preparing a listing of acronyms. This list, as reviewed and approved by the Project Team, is also included in this report.

The RP met from December 1, 2003 through December 5, 2003, in Columbia, MD. At the beginning of the meeting, the RP and the Project Team were introduced to the ASME/RSI peer review process. The introduction was followed by presentations by the Project Team and a discussion period. The agenda of this meeting is shown in the Appendix. At the end of the meeting, the RP met in an executive session to develop its strategy for writing the report. Subsequently, the RP wrote its findings and recommendations. The *Report of the Review Panel* was then copyedited.

Consistent with the procedures established by the ASME/RSI process, the copyedited *Report of the Review Panel* was provided to DOE for identification of potential errors, misunderstandings, and areas of ambiguity. The Technical Secretary contacted the members of the RP reporting the comments received from DOE which were considered by the RP.

Charles O. Velzy
A. Alan Moghissi

Appendix

ASME/RSI Peer Review
Spent Nuclear Fuel Canister Welding Concept

Columbia, MD - December 1–5, 2003

AGENDA

Monday, December 1, 2003

Hilton Columbia

8:00 a.m.	ASME/RSI Peer Review Process	A. Alan Moghissi, ASME/RSI
9:00 a.m.	Introduction • Overview of Application • Economics • Approach	A.D. Watkins, INEEL
10:00 a.m.	Break	
10:15 a.m.	Requirements • Fabrication • Radiation Requirements Matrix Process Selections/Code Requirements/ 10% Notch	A.D. Watkins
12:00 noon	Lunch	
1:00 p.m.	Welding Equipment • Requirements • Hardware	D.P. Pace, INEEL
2:30 p.m.	Break	
2:45 p.m.	Welding Equipment *(continued)* • Approach	
3:30 p.m.	Grinding Equipment • Requirements	D.P. Pace
4:00 p.m.	Break	
4:15 p.m.	Grinding Equipment *(continued)* • Hardware • Approach	
5:30 p.m.	Adjournment	

ASME/RSI Peer Review
Spent Nuclear Fuel Canister Welding Concept

Columbia, MD - December 1–5, 2003

AGENDA

Tuesday, December 2, 2003

Hilton Columbia

8:00 a.m.	NDE Overview Requirements Visual Examination • Requirements • Hardware • Approach	A.D. Watkins, INEEL
9:45 a.m.	Break	
10:00 a.m.	Eddy Current • Requirements • Approach • Hardware • Repair Groove Inspection Ultrasonic Inspection • Requirements • Approach • Hardware • Pass by Pass Inspection	T.R. McJunkin, INEEL
12:00 noon	Lunch	
1:00 p.m.	Executive Session	
2:30 p.m.	Break	
2:45 p.m.	Discussion	
4:30 p.m.	Executive Session	
5:30 p.m.	Adjournment	

ASME/RSI Peer Review
Spent Nuclear Fuel Canister Welding Concept

Columbia, MD - December 1–5, 2003

AGENDA

Wednesday, December 3, 2003

Institute for Regulatory Science

8:00 a.m.   Executive Session

5:00 p.m.   Adjournment

Thursday, December 4, 2003

Institute for Regulatory Science

8:00 a.m.   Executive Session

5:00 p.m.   Adjournment

Friday, December 5, 2003

Institute for Regulatory Science

TBD   Closeout Session

Executive Summary

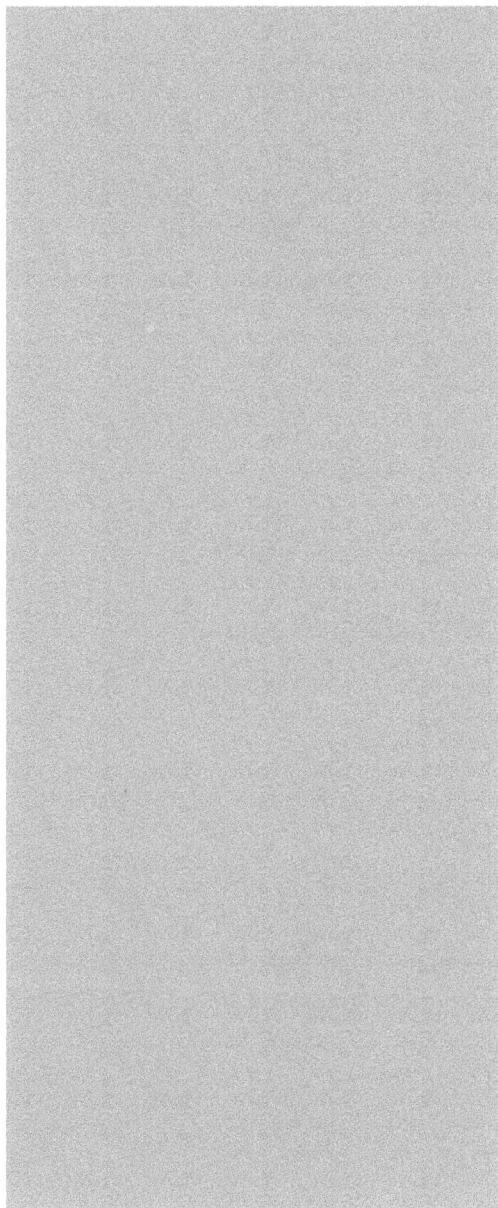

The National Spent Nuclear Fuel Program (NSNFP) developed standardized U.S. Department of Energy (DOE) SNF canisters for the handling and interim storage of SNF at various DOE sites, as well as the transport, handling, and disposal of SNF at the repository. The standardized canisters for DOE SNF have a 457 mm (18 in) or 610 mm (24 in) outside diameter (OD); are fabricated from type 316L stainless steel (SS); and are designed to go in the center of a repository waste package.

Seal welding of the standardized canister must be performed after DOE SNF is loaded into the containers. As such, the welding and nondestructive examination (NDE) will be performed remotely in a hot cell. Because the containers may be part of a storage, transport, and/or disposal system during their useful lifetime, it is the intent to have the vessels designed and built to ASME Section III, Division 3 requirements.

The welding, inspection, and repair equipment are stationed on a carrousel that rotates around the SNF-standardized canister. All power supply and electronics are housed behind a shielding device to minimize the equipment exposure to a high-radiation environment. Each of the three towers of the carrousel supports one component (welding, inspection, and repair) of the system. Only the end effector portion of the equipment is exposed to a high-radiation environment. Iterative grind and inspect steps will be performed to verify the removal of a defect. All repair cavities will have a consistent geometry.

The design and execution of the spent nuclear fuel (SNF) canister welding concept are consistent with established scientific and engineering principles and standards. The Project Team (PT) appears to be aware of relevant scientific and engineering data on remote welding for canisters, but did not cite specific publications and references. The PT did not address the details of the technical drawings and the adequacy of the tolerances on the quality of the welds.

Based on the extent of prior use, research, development, and application of the proposed welding process, the PT has demonstrated the appropriateness of the approach. In general, the Project Team has demonstrated sufficient understanding of the baseline technology and specific component requirements. However, the PT did not present the requirements for the relationships among the different components; between the system and the hot cell; and between the system and the operator.

The PT provided a description of a suggested set of NDE testing procedures, but did not evaluate their performance in a harsh environment. Both the proposed eddy current and phased array ultrasonic testing techniques need to be evaluated under operating conditions for signal/noise ratio, repeatability, and durability (transducers and cabling). Other possibilities, such as non-contact ultrasonic techniques and laser-profiling systems, were discussed in a limited way. These techniques may be helpful in case the water couplant problem for the ultrasonic technique is insurmountable due to temperature problems. Although the PT has considered validation procedures for their NDE testing, suggestions for the content of a comprehensive validation plan were not presented.

The initial focus was shown for components and process tools. Relatively little was presented regarding the integration of these components into an overall conceptual design which includes interfaces (among the different components; between the system and the hot cell; and between the system and the operator); system architecture; and performance requirements.

The PT has adequately identified and addressed the environmental—including human health—risks for the purposes of a conceptual design. However, the PT has not yet estimated the residual contamination level, and the frequency and time required for human involvement in operations, maintenance, and repair internal to the hot cell, which are necessary for the final design of the chosen concept.

The conceptual design of the welding, inspection, and repair system does have the potential to meet or exceed the customer requirements. It is premature to consider optimization of a conceptual design. The PT has recognized that there are conflicting requirements that are outside it's purview.

The PT has demonstrated good design principles in its approach and awareness of the problems related to the canister welding concept. The

PT is capable to address the hurdles encountered in the implementation of this conceptual design.

There is a significant discrepancy between the temperatures provided in the specification for the DOE SNF canister and the temperature assumed in the conceptual design.

This project cannot be performed in isolation, independent of other activities. Successful completion is dependent upon the effective coordination among the different groups involved in the formal design.

Based on its technical merit, there is a reasonably high likelihood for the successful implementation of this project. This project should be continued in the development of a detailed design.

Based on a careful assessment of the information presented to the Review Panel (RP) and the findings developed in response to the review criteria, the RP provides the following recommendations:

1. This project should be continued provided the remainder of these recommendations is seriously considered.
2. An organizational structure should be created in order to ensure an appropriate coordination among the different groups involved in the formal design of the welding, inspection, and repair system; hot cell facility; SNF canister; and packaging and transportation platform.
3. The PT must resolve the discrepancy between the temperatures provided in the specification for the DOE SNF canister and the temperature assumed in the conceptual design.
4. The PT must define geometric dimensioning and tolerances (GD&T) for the head and canister assembly to ensure acceptable quality welds.
5. The PT must investigate the potential fractures of the canister seal weld due to dynamic loads experienced during transportation and handling. The applicability of codes and standards dealing with dynamic loading should be assessed.
6. The PT should evaluate alternative designs to the backing ring that do not increase the potential for cracking and/or corrosion in the welded zone.

7. The proposed nondestructive examination (NDE) eddy current and ultrasonic techniques should be tested and evaluated under operating conditions of radiation and temperature to determine the feasibility of the proposed approaches.

8. The PT should evaluate in greater detail non-contact ultrasonic techniques such as laser ultrasonics; electromagnetic acoustic transducers (EMATs); and air-coupled ultrasonics for the inspection process.

9. The PT should develop a validation plan sufficient to qualify the selected NDE techniques for operation in a harsh environment with high reliability.

10. The PT should develop a formal specification that includes performance requirements; system architecture; and interfaces (among the different components; between the system and the hot cell; and between the system and the operator).

11. The PT should estimate the residual contamination level, and the frequency and time required for human involvement in operations, maintenance, and repair internal to the hot cell.

12. In a subsequent design, the PT should develop a plan for automatic motion control; calibration; tool changeout; and coordination of multiple axes positioning equipment in the execution of welding, inspection, maintenance, and repair operations.

Peer Review
Process

INTRODUCTION

There is consensus within the technical community on the definition, process, and key criteria for the acceptability of peer review. Peer review consists of a critical evaluation of a topic by individuals who—by virtue of their education, experience, and acquired knowledge—are qualified to be peers of an investigator engaged in a study. A peer is an individual who is able to perform the project, or the segment of the project that is being reviewed, with little or no additional training or learning.

Recognizing that peer review constitutes the core of acceptability of scientific and engineering information, virtually all professional societies of scientists and engineers have instituted formal procedures for peer review for their activities. The American Society of Mechanical Engineers (ASME), also known as ASME International, has over a century of experience in peer review. Consistent with its mission and tradition, ASME, in cooperation with the Institute for Regulatory Science (RSI), has established a peer review program devoted to the review of activities of various government agencies (ASME 2003, RSI 2003). The reports of the peer reviews resulting from this program have been published (ASME/RSI 1997, 1998, 1999, 2000, 2001a, 2001b, 2001c, 2002a, 2002b, 2002c, 2002d, 2003a, 2003b, 2003c).

PEER REVIEW PROCESS

The structure of the peer review process established by the ASME/RSI team consists of a tiered system. For each specific area, the entire process is overseen by a committee. The review of specific topics is performed by Review Panels (RPs).

Peer Review Committee for Energy and the Environment

The Peer Review Committee for Energy and the Environment (PRCEE) is a standing committee of ASME formed to oversee peer review for one particular program in an agency. Its members are chosen on the basis of their education, experience, peer recognition, and contribution to their respective areas of competency. An attempt is made to ensure that all needed technical competencies and diversity of technical views are represented in the PRCEE. The members of the PRCEE must be approved by the Board on Research and Technology Development of the Council on Engineering of ASME. The PRCEE includes an Executive Panel (EP) that is responsible for the day-to-day operations of the PRCEE. Except for the EP, membership in ASME is not required for appointment to the PRCEE. As the overseer of the entire peer review process, the PRCEE enforces all relevant ASME policies, including compliance with professional and ethical requirements. A key function of the PRCEE is the approval of the appointment of members of RPs for a specific project.

Review Panels

The review of a project, a document, a technology, or a program is performed by an RP consisting of a small group of highly-knowledgeable individuals. Upon the completion of their task, the RPs are disbanded. The selection of reviewers is based on the competencies required for the specific review assignment. The number of individuals in an RP depends upon the complexity of the subject to be reviewed. The selection of a reviewer is based on the totality of that individual's qualifications. However, there are several generally-recognized and fundamental criteria for assessing qualifications of a reviewer. These are as follows:

1. **Education:** A minimum of a B.S. degree, preferably an advanced degree in an engineering or scientific field, is required for any peer reviewer.

2. **Experience:** In addition to education, the reviewer must have significant experience in the area that is being reviewed.

3. **Peer recognition:** Election to an office of a professional society, serving on technical committees of scholarly organizations, and similar activities are considered to be a demonstration of peer recognition.

4. **Contributions to the profession:** Contributions to the profession may be demonstrated by publications in peer-reviewed journals. In addition, patents, presentations at meetings where the papers were peer-reviewed, and similar activities are considered to be contributions to the profession.

5. **Conflict of Interest:** One of the most complex and contested issues in peer review is a set of subjects collectively called conflict of interest. The ideal reviewer is an individual who is intimately familiar with the subject and yet has no monetary interest in it. Despite this apparent difficulty, the ASME and similar organizations have successfully performed peer review without having a real or apparent conflict of interest. The guiding principle for conflict of interest is as follows: *Those who have a stake in the outcome of the review may not act as a reviewer or participate in the selection of reviewers.*

Due to the multidisciplinary nature of many projects reviewed by the ASME/RSI team, rapid identification of qualified peer reviewers and their availability to participate in the review process are key ingredients for a successful program. The process used for the identification of reviewers is multifaceted. The Administrative Manager of the Peer Review Program receives recommendations from sources within ASME; previous members of the RP; sister societies; other organizations and individuals; the U.S. Department of Energy (DOE); DOE contractors; and others. However, the selection of peer reviewers is based entirely on criteria identified by ASME. The details of various aspects of peer review, including conflict of interest, can be found in the ASME *Manual for Peer Review* (ASME 2003) and the associated procedures (RSI 2003).

COOPERATION WITH OTHER PROFESSIONAL SOCIETIES

The ASME is a large professional engineering society having in excess of 125,000 members. Although the predominant discipline of the members is mechanical engineering, there are members who—by virtue of their education, training, or experience—are competent in other disciplines. The Council on Engineering includes divisions ranging from classical mechanical engineering (design, heat transfer, and power) to solar engineering; environmental engineering; and safety and risk analysis. Despite the diverse competency within ASME, it is recognized that on occasion it will become necessary to peer review activities which include disciplines that are outside the areas of competency of ASME and its members. These disciplines may include geology, hydrology, toxicology, and ecology. Consequently, ASME has reached formal and informal agreements with its sister societies to identify qualified reviewers in areas outside of those covered by the membership of ASME.

PERFORMING ORGANIZATIONS

The Center for Research and Technology Development of ASME manages a number of scientific and engineering activities, including peer reviews. Because of ASME's conscious effort to maintain a small in-house staff, it relies upon other organizations to provide detailed project management services in its research, development, and similar activities. Accordingly, ASME and RSI joined forces in a collaborative effort to perform the peer review for the U.S. Department of Energy. While the ASME staff in Washington, DC provides the staff support for the PRCEE, the detailed management and staff support for the RPs is provided by RSI.

American Society of Mechanical Engineers

As one of the largest professional engineering societies, ASME has a long and distinguished history. Its activities are carried out primarily by members who volunteer their time in support of engineering and scientific advancement. For obvious reasons, ASME also has a paid staff to manage the day-to-day operations of such a large professional society. ASME has a detailed structure for its operation, consisting of councils, boards, divisions, and committees. The Council on Engineering has 38 divisions, including: Environmental Engineering; Solid Waste Processing; Nuclear Engineering; Safety Engineering; and Risk Analysis. The Council on Codes and Standards develops ASME codes and standards that are the backbone of many industries—including power production—worldwide. The Council on Codes and Standards is also responsible for the development of standards for activities such as certification of incinerator operators. The ASME was a founding member of the American Association of Engineering Societies and a founding member of the American National Standards Institute.

Institute for Regulatory Science

RSI is a not-for-profit organization chartered under section 501(c)3 of the Internal Revenue Service. It is dedicated to the idea that societal decisions must be based on the best available scientific and engineering information. According to the RSI mission statement, peer review is the foundation of the best available scientific and engineering information. Consequently, RSI has promoted peer review within government and industry as the single most important measure of reliability of scientific and engineering information. In its activities, RSI seeks the cooperation of scholarly organizations. Historically, a large number of RSI activities have been performed in cooperation with ASME. RSI is located in the Washington, DC metropolitan area.

Project Summary

PRELIMINARY DESIGN SPECIFICATION FOR STANDARDIZED SPENT NUCLEAR FUEL CANISTERS OF U.S. DEPARTMENT OF ENERGY

INTRODUCTION

The Nuclear Waste Policy Act (NWPA 1982) assigned U.S. Department of Energy (DOE) the responsibility for managing the disposal of spent nuclear fuel (SNF) and high-level waste (HLW) of domestic origin. The Nuclear Waste Policy Amendments Act (NWPAA 1987) designated Yucca Mountain, NV as the candidate site for the geologic repository. Furthermore, the development of general-applicable environmental standards and the licensing of the repository are the responsibility of the U.S. Environmental Protection Agency (USEPA) and the U.S. Nuclear Regulatory Commission (USNRC) respectively. These laws are also responsible for the disposal of SNF and HLW, including the development and licensing of a geologic repository to DOE's Office of Civilian Radioactive Waste Management (OCRWM). Additionally, a Presidential Memorandum (1985) stated that there was no compelling reason to build a separate repository for defense HLW; therefore, that waste will also be disposed in the civilian geologic repository.

The NWPA (1982) defines SNF as the fuel that has been withdrawn from a nuclear reactor following irradiation, the constituent elements of which have not been separated by reprocessing. As used in the design specification, SNF is also defined to include non-fuel components as identified in 10 CFR Part 961, Appendix E (DOE 2003). In the NWPA (1982), HLW is defined as the highly-radioactive material resulting from reprocessing SNF, including liquid waste produced directly in reprocessing; any solid material derived from such liquid waste that contains fission products in sufficient concentrations; and other highly-radioactive material that has been determined by the USNRC, consistent with law, to require permanent isolation. As used in the design specification, HLW is defined to include solidified HLW, resulting from commercial and defense operations and SNF. Consequently, the repository will accept SNF and solidified HLW.

The NWPA (1982) declared storage of SNF in monitored retrievable storage (MRS) facilities as a safe and reliable option for managing SNF. In the NWPAA (1987), Congress authorized the Secretary of Energy to site, construct, and operate one MRS facility. As stated in 10 CFR Part 72 (USNRC 2003c), the MRS will have an initial 40-year license term with the option for renewal by the USNRC. The MRS, if it is built, will provide temporary storage for SNF until the SNF is shipped to the geologic repository for permanent disposal. Additionally, SNF can pass through or flow through the MRS to the geologic repository.

Consistent with the requirements of the NWPA, including its amendments, the National Spent Nuclear Fuel Program (NSNFP) of DOE is developing a set of standard canisters for (SNF), including its handling, interim storage, transportation, and disposal in the national repository. The mission of this program is to safely, reliably, and efficiently manage the DOE-owned SNF, have the SNF returned to the U.S. from foreign research reactors (FRR), and prepare it for disposal. This activity is cooperatively performed with the OCRWM; the Idaho National Engineering and Environmental Laboratory (INEEL); DOE Hanford Site; Oak Ridge National Laboratory; Argonne National Laboratory; and the Savannah River Site (SRS).

Within DOE, the Office of Environmental Management (EM) is responsible for the interim management and preparation for disposal of the DOE SNF (DOE 1999), which includes SNF generated from activities related to nuclear weapons production, research programs, and others. EM establishes the methods to be employed in the treatment, handling, storage, and preparation for disposal of the DOE SNF, including both current inventory and expected receipts. The NSNFP, operating from the INEEL, assists EM in implementing these methods.

Consistent with its mission, the NSNFP has developed a specification to provide the requirements and necessary information for designing the standardized canisters to be used for handling, interim storage, transportation, and disposal in the national repository of DOE SNF

(DOE 1999). This design specification addresses two different outer diameter (OD) sizes of DOE SNF standardized canisters, including two different lengths for each canister OD, resulting in a total of four unique canister geometries. This design specification does not consider using these DOE SNF canisters for either U.S. Navy or commercial SNF or HLW (either commercial or defense) materials. Similarly, it does not provide any procurement-specific information.

Although using the same terminology, the standardized DOE SNF canister should not be confused with the containers used in current interim SNF storage systems for commercial SNFs. Many independent spent fuel storage installations (ISFSI) or dry storage system vendors also call their SNF container a canister. However, the commercial nuclear industry storage canister (hereafter referred to as the storage industry canister) is typically 5 to 6 feet (1.5 to 1.8 m) in diameter. In contrast, the DOE SNF canisters are approximately 1.5 to 2 feet (0.45 to 0.60 m) in diameter.

The American National Standard Institute (ANSI) provides standards for a variety of activities, equipment and practices. In the Foreword of one of its standards, ANSI (1992) states: "...the safe storage of spent fuel assemblies is achieved by maintaining a minimum of two independent barriers between the fuel and the environs. The fuel cladding is considered the primary barrier for undamaged fuel. [...] In addition, the complete confinement system for the stored fuel is conservatively designed to withstand damaging events [...] so that there is an effective secondary barrier(s) to the release of radioactive materials under all credible conditions." The storage industry canister is intended to provide the secondary barrier for the commercial SNF. In contrast to industry canisters, the DOE SNF canisters are intended to constitute the primary barrier or cladding replacement for interim storage and transportation to the repository. Consequently, not only damaged, intact, or failed DOE SNF are expected to be placed directly into the standardized DOE SNF canisters. The design requirements for DOE SNF canisters are listed in Table 1.

Table 1. Design requirements matrix for SNF canisters (DOE 1999). Note that this table does not specifically address quality assurance requirements.

Requirements	Subject
Federal Regulations	
10 CFR Part 60.131 (h)	Criticality Control
10 CFR Part 60.135 (a) (1)	In situ chemical, physical, and nuclear properties do not compromise function of waste package
10 CFR Part 60.135 (a) (2)	Design considerations for waste package and content interactions
10 CFR Part 60.135 (b) (1)	Waste packages and its components shall not contain explosive, pyrophoric, or chemically reactive material
10 CFR Part 60.135 (c) (1)	DOE SNF shall be in solid form and placed in sealed containers
10 CFR Part 71.63 (b)	Double containment of certain SNF
10 CFR Part 72.122 (h) (1) & (5)	Confinement of SNF
10 CFR Part 72.122 (1)	Retrievability from storage
10 CFR Part 72.124	Criticality control
10 CFR Part 72.128 (a) (3)	Confinement during storage
DOE OCRWM Documents	
Disposability Interface Specification – July 1998 (Draft)	
Disposability Standard 2.1.20	Canister shell and lid materials shall be low-carbon austenitic stainless steel or stabilized austenitic stainless steel or other equally corrosion-resistant alloys
Disposability Standard 2.1.21	Canisters shall be vacuum dried, backfilled with an inert gas (e.g., helium) and sealed
Disposability Standard 2.1.26	Canisters shall have a unique alphanumeric identifier
Disposability Standard 2.1.27	Canisters not seal-welded shall have a tamper-indicating device
Disposability Standard 2.1.28	Damage or deformation to canisters shall be limited such that the canisters can (1) still be lifted and moved, (2) continue to meet the dimensional envelope required for disposal (loading into the disposal container), and (3) maintain a seal
Disposability Standard 2.2.20.3	Canisters shall have the capability to stand upright without support on a flat surface and be capable of being placed, without forcing, into a right-circular cylindrical cavity of the proper dimension
Disposability Standard 2.2.21.3	Canisters shall not exceed the total weight limits as specified (DIS conflicts with ICD values but ICD values are used)

Table 1. (cont'd).

Requirements	Subject
DOE OCRWM Documents	
Disposability Interface Specification – July 1998 (Draft)	
Disposability Standard 2.2.22.2	Canisters shall be capable of being lifted vertically with remote handling fixtures
Disposability Standard 2.3.22	Canistered SNF shall be shown to have a calculated keff of 0.95 or less
Disposability Standard 2.4.21	Canisters shall comply with thermal output limits [currently vague on DOE SNF canisters]
Disposability Standard 2.4.22	Canisters shall comply with surface contamination limits
Disposability Standard 2.4.23	Canisters shall not exceed pressure limits
Disposability Standard 2.4.24	Canisters shall have no detectable leak rate at the time of receipt at the repository. At a minimum, the canister shall be leak tested using OCRWM-approved method or shown via an OCRWM-approved method of fabrication controls and volumetric inspections to be properly sealed. The canister shall be reevaluated prior to shipment, as required, if suspected of leaking.
Interface Control Document – March 1999	
ICD Section 10.1.1	Canisters are right-circular cylinders and after being filled with SNF can stand vertically on a flat surface
ICD Section 10.1.2	Canisters shall not exceed weight limits
NUREG-1617 (Draft) – March 1998	
Section 4.5.1.3	Packages designed for the transport of damaged SNF include packaging of the damaged fuel in a separate inner container (second containment system) that meets the requirements of 10 CFR 71.63 (b)

FUNCTIONAL REQUIREMENTS

Project requester and end user

The development of the standardized U.S. Department of Energy (DOE) spent nuclear fuel (SNF) canisters is guided under the direction of the National Spent Nuclear Fuel Program (NSNFP) as an implementor of the DOE SNF Program. The DOE SNF Program is the project requestor.

The NSNFP personnel is located at the Idaho National Engineering and Environmental Laboratory (INEEL) administered by the Lockheed Martin Idaho Technologies Company (LMITCO). The end user of the finished canisters will ultimately be the geologic repository, where SNF will be permanently disposed. However, where applicable, the fuel custodians at the DOE sites (Hanford, INEEL, and SRS) will be responsible for loading, handling, and storing DOE SNF in these standardized canisters.

Purpose of the standardized canisters

The purpose of a standardized canister for DOE-owned SNF is as follows:

1. To provide an easy and standard handleable unit to confine DOE SNF materials
2. To be durable for storing SNF
3. To provide easily-transportable units
4. To be a unit for final disposal at the national repository, without the necessity of the DOE SNF being removed from the canister or re-opening a sealed canister.

The standardized DOE SNF canisters will provide long-term advantages in four basic areas:

1. Handling and safety:
1.1. Ensure that the DOE SNF is directly handled (bare) only once during its interim storage, transportation, and final disposal handling stages
1.2. Promote standardized handling of materials at all facilities, interim storage sites, and the repository
1.3. Improve human factors and performance efficiency of handling procedures
1.4. Provide contamination control to reduce safety risks
1.5. Reduce the risk of radiation exposure to the workers and the public by: 1) not opening the canisters during inspection, by 2) providing better automated, remote handling possibilities

2. Storage and disposal:
2.1. Provide standardized storage configurations at the interim storage sites and the repository
2.2. Minimize the number of canisters which must be qualified for interim storage and receipt by the repository

3. Transportation:
3.1. Promote standard sizing configurations to fit into transportation casks or packagings for efficient packaging design and, consequently, more efficient loading and unloading procedures
3.2. Promote standardized sizes and compatibility for maximum utilization of transportation facilities and equipment

4. Economics:
4.1. Reduce overall handling, transportation, and repository emplacement costs in a simpler, more integrated operation, since fewer modifications will be required for The Office of Civilian Radioactive Waste Management (OCRWM) to accept DOE SNF
4.2. Minimize system costs by reducing the design costs of multiple canister designs

General functional criteria

Figure 1 schematically shows the various paths that the current DOE SNF may follow from the point it begins its initial disposal handling until placement into the repository. This basic structure shows the route SNF can take from initial handling through interim storage to repository disposal. It can be used as a guide in the demonstration of the technical adequacy of the canister design as well as demonstrating compliance with regulatory requirements. The indicated process guides the general functional criteria of the standardized canisters. These functional criteria are:

1. The canisters shall be right circular cylinders.
2. Two different diameter-sized canisters shall be designed, with nominal outer diameters (OD) of 24.00 inches (610 mm) and 18.00 inches (457 mm), to accept a significant portion of the various DOE SNF currently in existence.

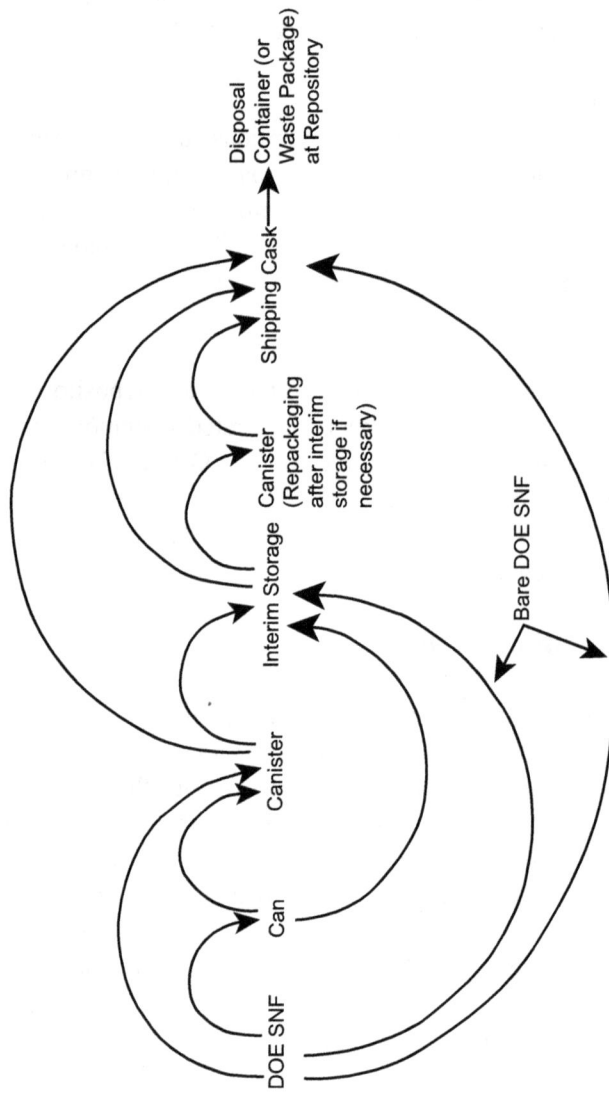

Fig. 1. Possible DOE SNF paths to repository disposal (DOE 1999).

3. Two overall lengths for each canister size, 3,000 mm (118.11 inches) and 4,570 mm (179.92 inches), will be considered maximum lengths from end to end, inclusive of the cap ends, labeling, and any handling fixtures.

4. The 610 mm (24.00 inches) OD canister can be accommodated into a repository waste package as a replacement for one of the HLW canisters, providing a canister for large-sized DOE SNF where needed, and to also provide a potential overpack option for the 457 mm (18.00 inches) OD canister.

5. The 457 mm (18.00 inches) OD canister can be accommodated in the center hole of a five-pack waste package at the repository as well as being able to be stored in various facilities at the INEEL, Hanford, or SRS.

6. The canisters must perform as required while subjected to the most severely anticipated environmental conditions and natural phenomenon postulated to occur for the entire service life of the canister.

7. The design of the standardized canisters shall be robust enough to accommodate the grappling and handling equipment configurations at the interfacing facilities when loaded to the weight limits of the canister.

8. Sealing of the canisters shall be accomplished by welding.

9. The canisters shall be designed to provide safe storage (in coordination with the storage facility design) of SNF at any location in the continental United States for a minimum of 40 years. The canisters—in coordination with the handling systems; interim storage systems; transportation systems of many different facilities; and the repository disposal system—must provide confinement for the SNF under all anticipated normal, off-normal, and accident conditions.

10. A safety analysis will need to be performed commensurate with the potential consequences of any activity being performed in conjunction with these canisters.

Figure 2 illustrates the various anticipated uses of the DOE SNF canister.

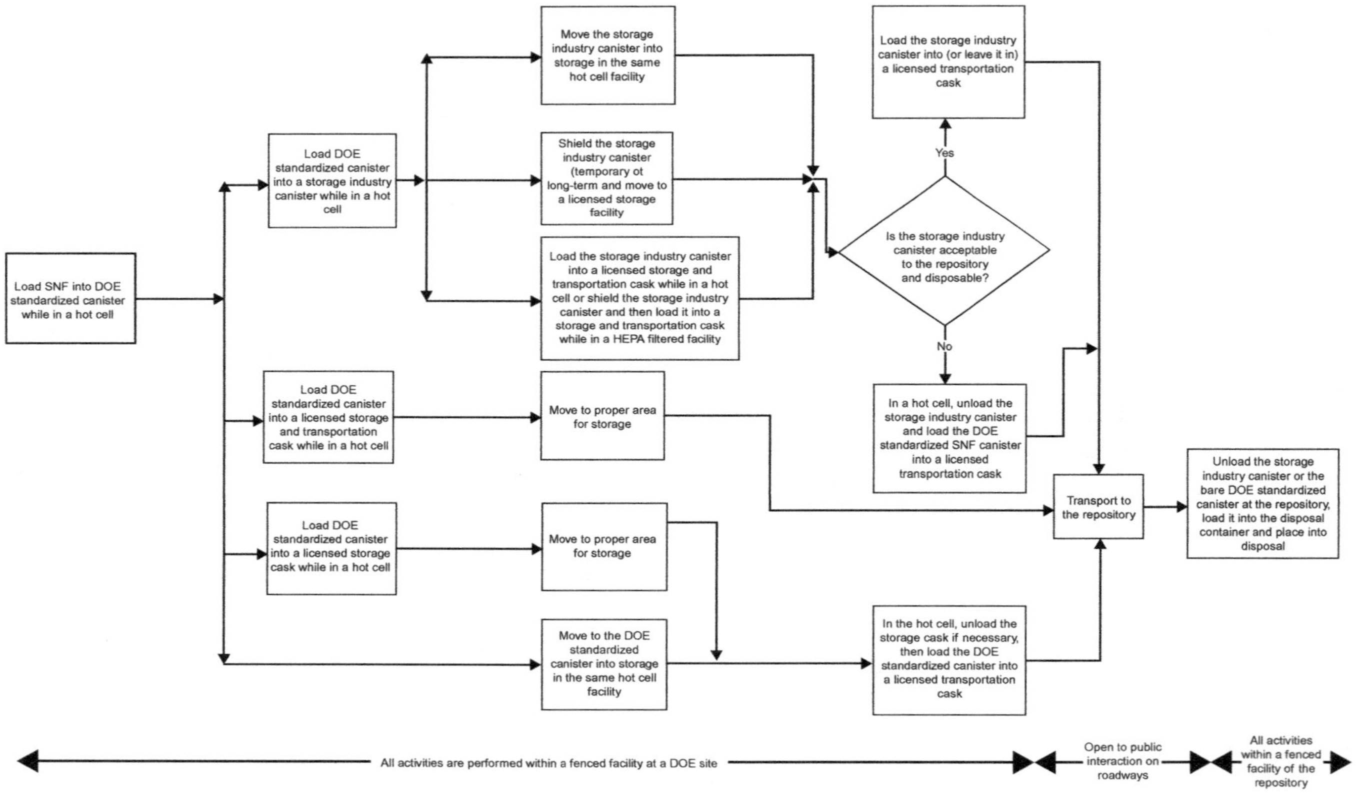

Fig. 2. Standardized DOE SNF canister usage (DOE 1999).

38

Assumptions for standardized canister design

The basic assumptions used for the functional performance of the standardized canister designs for DOE-owned SNF are:

1. The standardized canisters shall have an inherently robust design such that they can be readily incorporated into storage and transportation systems meeting 10 CFR Part 72 (USNRC 2003c) and 10 CFR Part 71 (USNRC 2003b) requirements respectively.
2. The standardized canisters shall be compatible with the currently known repository requirements.
3. The design of any internal components (such as baskets, spacers, sleeves, dividers, and cans), necessary for the loading of the SNF and for the control of criticality, must be constrained by the existing design and interior dimensions of the DOE SNF canisters.
4. The handling facilities will have to accommodate modification to their facilities if these canisters do not fit within the safety envelope of: 1) the design criteria for accidental drops, and 2) confinement for expected accidental release scenarios.
5. The SNF custodian shall select those canisters, or combinations of canisters, that provide an optimum packing configuration for the custodian's SNF.
6. The DOE SNF canisters shall be capable of accepting intact, failed, or damaged SNF, directly or canned.
7. The DOE SNF is loaded into a canister in a hot cell, or shielded facility, so that the fuel can be properly dried and adequate radiation and temperature measurements can be taken to ensure compliance with applicable canister design limits. If necessary, the DOE SNF can be initially placed into a canister underwater (drain holes are provided in the canister design), but the fuel and canister must then be properly dried. Final loading and testing would still occur in a hot cell or shielded facility.
8. The loading of the SNF into a canister will not cause significant localized thermal gradients in the pressure boundary, nor will it result in significant bowing concerns for the canister.
9. After SNF loading and radiation and temperature acceptance, the canisters shall be backfilled with an inert gas (e.g., helium). Canister sealing during storage is optional (depending on the storage system

used) but seal-welded canisters (backfilled with an inert gas) are required for transportation and repository disposal.

10. The amount of added pressurization from the SNF, during interim storage and transportation, is assumed to be small, such that the actual maximum pressure experienced by the canisters during their containment lifetime is less than the repository acceptance pressure of 151.7 kPa (22 psig)

11. After loading the canisters with SNF, all canisters will require either remote-handling due to a lack of canister shielding, or placement into another container or overpack which provides adequate shielding.

12. Inspections of the SNF after interim storage and prior to transportation to the repository are not anticipated at this time.

The SNF being considered for movement to interim storage, transportation, and final disposal are those listed in the Spent Fuel Database, currently maintained by the NSNFP. The two standardized sized canisters—457 mm (18.00 inches) and 610 mm (24.00 inches) nominal outer diameter (OD), each with two suggested lengths as listed in the specification—shall provide confinement for the listed DOE SNF, with minimal exceptions for larger sized SNF that may or may not have the ability to be transported as bare SNF in transportation casks. Some SNF or other radioactive materials, such as the particulate waste described in 10 CFR 60.135(c)(2) (USNRC 2003a), may require an interior sealed container. An example of such a container is a standard high-integrity can (SHIC) with approximately 5-inch (12.5 cm) OD, which provides for ease of packaging due to their smaller size. They also maintain the structural integrity of the SNF or radioactive material; reduce the possibility of criticality configurations; reduce the possibility of occurrence of other dangerous scenarios; and reduce potential gas generation concerns. The SNF that is placed in the canister will be assumed to be adequately characterized for storage, transportation, and final disposal when sealed into the canister. This information includes radiation shielding, decay heat removal, corrosion, gas generation, criticality control, and other parameters necessary for adequate safety.

Conformance with requirements

The basic design must conform to the following applicable sections of these higher-level documents and organization specific criteria:

1. Requirements 10 CFR Parts 60, 71, and 72 (USNRC 2003 a,b,c)
2. Requirements of USNRC for interim storage and transportation to the repository
3. Department of Transportation (DOT) requirements for transportation to the repository
4. Requirements of DOE-OCRWM for transportation, handling, and disposal at the repository
5. Compliance with applicable voluntary standards such as Codes and Standards of the American Society of Mechanical Engineers (ASME)

Necessary analyses

A design analysis shall be required to clearly demonstrate compliance with the specified design criteria. The Design Report—certified by a competent and qualified professional engineer associated with an N-stamp holder as described in Codes and Standards of ASME—must be submitted by the designer of the DOE SNF canisters. A safety analysis shall be required to be commensurate with the potential consequences of any activity being performed in conjunction with the use of the DOE SNF canisters. This safety analysis must be performed by those SNF custodians using the DOE SNF canisters. SNF dimensions and composition must be characterized for interim storage, transportation, and disposal requirements including radiation shielding, decay heat removal, and criticality control. The DOE SNF must be characterized for repository disposal performance assessment in support of licensing requirements which have not yet been finalized.

Prerequisites for the design specification for the DOE SNF canisters

This preliminary design specification establishes a common basis for all standardized DOE SNF canisters so that they can be accepted for disposal at the repository. The design specifications must consider that the DOE SNF canister's final design must be integrated with the total system design, including storage and transportation requirements. Different groups of DOE SNF canisters may be stored in a variety of facilities, and these same groups of DOE SNF canisters may even be transported in a variety of transportation casks. Each unique group of DOE SNF canisters will require Design Specifications and Design Reports as specified by ASME (2001). The Design Reports must establish the design bases and the specified uses of the DOE SNF canister. The Design Reports may also have to be updated as necessary when design information changes. The preliminary design specification assumes that after the DOE SNF is placed in the canister, their use shall be limited to that indicated in four areas as follows:

1. Initial loading and canister handling: Prior to loading, all canister materials shall be suitably examined, and all canister welds necessary for the structural integrity of the pressure boundary shall be volumetrically examined, using either radiography (RT) or ultrasonic (UT) methods to assure weld integrity. During SNF loading, due to the potentially high levels of radiation, all final loading operations, sealing, and testing must be performed remotely. Therefore, it is assumed that these activities will be performed while the canisters are inside of a hot cell or shielded facility. After loading the SNF and any necessary internals, the canister top head will be attached. A clamping device (e.g., a dearman clamp) may be necessary to minimize ovalization of the canister shell for proper head fit-up. The top head closure weld (a simple butt weld) will be made using a vessel head that has a backing ring. The backing ring will help with weld fit-up and also protect the canister contents during welding. Inspection of the top head weld can be achieved using UT techniques.

Sealing of the DOE SNF canisters may be required for interim storage, depending on the type of storage system utilized. However, if the storage system permits it, the DOE SNF canisters may be sealed after interim storage. Incorporated into the DOE SNF canister design is the option of a threaded plug in the top and bottom head. However, prior to transportation and repository disposal, the DOE SNF canister must be seal-welded closed. The DOE SNF canisters shall be backfilled with an inert cover gas (e.g., helium) inside of the canister at a pressure of 13.8 to 27.6 kPa (2 to 4 psig). The final closure weld shall implement a welding procedure that can be qualified to yield leak-tight welds. A leak-tight weld shall be considered equal to or better than the required leak rate necessary to satisfy requirements of 10 CFR Parts 71 (USNRC 2003b) and 72 (USNRC 2003c). At a minimum, in order to demonstrate compliance with the ASME (2001) Code, the DOE SNF canister shall be helium leak-tested as identified in Section V, Article 10, Appendix IV of ASME Code to verify that no leakage is detected that exceeds the rate of 10^{-4} std cm^3/s.

Once loaded with SNF, the DOE SNF canisters will have to be handled a number of times. These operations include, but are not necessarily limited to: preparing the canister for interim storage; preparing the canister for transportation; and preparing the canister for disposal at the repository. All of these handling situations are assumed to be performed within the confines of a facility that has high-efficiency particulate air (HEPA) filter capabilities. With this assumption, design requirements associated with accident events are significantly reduced. An accidental drop of a DOE SNF canister (by itself) may not result in exceeding the facility's offsite dose limits.

With this approach, the DOE SNF canister is not required to satisfy any specific ASME (2001) Boiler and Pressure Vessel Code design stress-limits for an accidental drop inside of a HEPA-filtered facility. Anytime the canister is outside of a HEPA-filtered facility, it will be inside of another containment vessel (a storage industry canister for interim storage or a transportation cask for transportation to the repository). This no-drop design philosophy for handling scenarios inside of a HEPA-filtered facility is similar to that followed by many SNF storage system vendors for their storage industry canisters at commercial nuclear power plants.

2. Interim storage: The DOE SNF canisters are expected to be used for interim storage (in conjunction with an acceptable storage system) at different locations and facilities throughout the U.S. before disposal at the repository. It is assumed that a SNF storage facility, properly designed and licensed in accordance with 10 CFR Part 72 (USNRC 2003c) will be utilized. It is also assumed that the DOE SNF canisters will be incorporated into those SNF storage systems. However, the specifics of any storage system are not currently known and may actually vary among the various DOE sites. Therefore, the precise details of how the DOE SNF canister is supported within the interim storage system and the associated storage industry canister details are simply not known at this time.

For the purposes of this design specification, the SNF storage system must be designed to incorporate the DOE SNF canister into its system. Because the canister is assumed to provide a robust design, it is anticipated that incorporation of the DOE SNF canister into any storage system will not be difficult. Many design considerations, including retrievability, seismic loads, accidental drop loads, and environmental conditions must be adequately addressed by the storage system vendor in order to ensure the proper care and use of these DOE SNF canisters.

A major assumption made, regarding the use of the DOE SNF canister in any SNF storage system, is that the DOE SNF canister will either stay in the hot cell facility for interim storage purposes or be placed inside of another storage container (either a storage industry canister, storage cask, or combination storage and transportation cask). Typically, this other container will be a containment vessel designed to withstand the anticipated operational and accidental loads identified for the SNF storage system.

3. Transportation: Although the DOE SNF canister is to be placed inside of a licensed transportation cask, the precise details of how the canister is supported within the cask and associated cask details are not known at this time. Therefore, for the purposes of this design specification, the transportation cask must be designed to incorporate the DOE SNF canister into its system. Because the canister is assumed to provide a robust design, it is anticipated that incorporation of the DOE SNF canister into a transportation cask system is achievable. Many design considerations, including the hypothetical accident conditions described in 10 CFR Part 71.73 (USNRC 2003b), must be adequately addressed by the entire transportation system (canister and cask) in order to ensure the proper care and use of these DOE SNF canisters.

4. Disposal at the repository: The loaded DOE SNF canisters will be handled at the repository during the unloading phase from the transportation cask and during the placement of the canister into the repository disposal container (or "waste package" once the final barrier weld is accepted). All of these activities are assumed to occur within the confines of a HEPA-filtered building. Placing the canisters into the waste package means that the DOE SNF canisters must be able to conform to the material compatibility and criticality issues indicated in 10 CFR Part 60 (USNRC 2003a). There are additional canister design requirements for the DOE SNF canister, but they are contained in various repository documents and cover allowable materials, pressures, weights, and other miscellaneous criteria, including labeling.

However, the DOE SNF canisters are not currently listed on the Q-list for repository equipment. This is an indication that the repository is not currently requiring the canisters to provide any safety function. Therefore, there are no quality assurance (QA) requirements that need to be imposed on the canister from the position of the repository.

Besides the design requirements for safely lifting and handling the DOE SNF canisters, the repository has indicated that the only remaining requirements imposed on the DOE SNF canisters are those associated with accidental drops or tip-overs of the waste package. Because the canister is assumed to provide a robust design, it is anticipated that incorporation of the DOE SNF canister into the waste package design accidental drop scenarios (2-meter drop and tip-over) should not be a difficult task. The repository personnel has indicated a requirement to remove the DOE SNF canister (not the individual fuel pieces) from a damaged waste package (due to tip-over or drop accident) so that it can be reloaded into another undamaged waste package. Once the waste package has been placed into a drift, the repository personnel has indicated that there are no additional design requirements imposed on the DOE SNF canister. This includes any requirements associated with the retrieval of a waste package as described in 10 CFR Part 60.111(b) (USNRC 2003a). The repository personnel is only anticipating the ability to retrieve a waste package, and is not concerned with removing any contents from a waste package during any retrieval efforts.

Limitations of uses of the DOE SNF canisters

By definition, intact SNF is fuel with cladding that has no hairline cracks or pinholes. Failed SNF is fuel with cladding that has hairline cracks or pinhole leak defects. Damaged fuel, by definition, is fuel with cladding that has defects greater than hairline cracks or pinhole leaks. The DOE SNF canisters must be designed to accommodate directly the placement of intact, failed, or damaged SNF inside of the canister. Using this approach, if the DOE SNF degrades during storage or transportation, the canister still satisfies the necessary requirements.

The DOE SNF canisters shall not be arbitrarily used for any other purposes beyond those indicated in the design specification without additional detailed analysis, evaluation, and testing in order to assure that other uses do not violate any of the limitations imposed by the design specification. Examples of use not considered herein include: 1) placing molten glass or molten HLW in these canisters; 2) dropping the canister onto sharp projectiles; and 3) dropping the canister from heights greater than that considered.

Canister barrier requirements

Because it has been assumed that damaged SNF will be loaded directly into the DOE SNF canister, the DOE SNF canister must perform the function of the cladding as the primary barrier. In this fashion, the storage system outer container (typically the storage industry canister), or the transportation cask, will continue to provide the function of the secondary barrier. Redundant sealing is achieved with this design approach. The design specification assumes that the storage or transportation systems do not provide the double barriers themselves.

Development of a robust design for the canister

Although details regarding SNF loading, interim storage, and transportation are not yet finalized, the National Spent Nuclear Fuel Program (NSNFP) is establishing a design for the DOE SNF canisters. This design must be robust, i.e., a design that has significant safety margins for the known or estimated loads. With this robust design, the NSNFP can proceed with the design and fabrication of the DOE SNF canisters. The storage system vendor and the transportation cask designer must incorporate the specified DOE SNF canister design into their respective system designs.

The robust design concept implies a canister developed to physically maintain containment proven by actual testing after a potential 9-m (30-foot) drop accident, similar to that required for transportation casks as described in 10 CFR Part 71.73 (USNRC 2003b). A robust design also minimizes canister deformations and damage that could potentially occur during normal, everyday handling scenarios, as described in the repository requirements.

Accidental drop scenarios should not be a major design problem for the interim storage vendor or the transportation designer. This is based on the assumption that the smaller DOE SNF canister will be placed inside another container consisting of the larger storage industry canister or the transportation cask. The storage industry canisters have already been successfully designed to accommodate their applicable site-specific and system-specific accidental drop or tip-over loads for commercial fuel. Transportation casks must be designed to accommodate 9-m (30-foot) drops onto an unyielding surface and still be able to remove the SNF contents. Therefore, the DOE SNF canister, being inside of the storage industry canister or transportation cask (and being robust), should present an easier design situation than deforming baskets. At most, impact limit-ers inside of the storage or transportation systems may be required for certain fuels with special criticality concerns.

The only remaining accidental drop concerns relate to when the canister is being handled by itself or is inside the waste package at the repository. As defined to date, these anticipated drop scenarios are: 1) a 7.3-m (24-foot) accidental drop of a DOE SNF canister at the repository while being placed into a disposal container; 2) a 2-m (6.56-foot) drop of the waste package with the DOE SNF canister inside; and 3) a tip-over of the waste package with the DOE SNF canister inside. These accident events are listed in order to provide a clear understanding of the intended use of the canister with regards to potential accidental drop events that have been identified.

There are significant advantages to constructing a DOE SNF canister that has a robust design and is drop resistant. By constructing a drop resistant canister, more assurances are provided that the canister can be readily incorporated into storage and transportation systems. These systems must be designed for certain postulated drop accidents in addition to other dynamic events. Also, if a drop accident occurs even in a HEPA-filtered facility, recovery would be much simpler and quicker for a canister that maintains containment of the SNF than if contamination of the facility resulted. It is believed that a severe drop accident resulting in contamination of the repository could shutdown the receiving operations for a significant period of time. The ripple effect from such a shutdown would impose nationwide schedule delays.

The design of a usable DOE SNF canister that satisfies ASME (2001) Boiler and Pressure Vessel Code Section III (Subsection NB, Level D or Division 3, Subsection WB) and hypothetical accident condition stress limits for an accidental drop scenario of significant height requires some form of energy absorption. Due to the high radiation levels involved, putting on and removing external impact limiters is not practical for the DOE SNF canisters. Consequently, having integral energy absorption capability is desirable. Given other constraints of size and useable volume, keeping the elastically calculated stresses below the acceptable stress limits for the DOE SNF canister, is extremely difficult even with self-contained energy absorption capabilities. Therefore, the NSNFP has designed the DOE SNF canister using a plastic strain approach. The canister is designed with symmetrical, skirted ends and a cylindrical shell sized to absorb a sufficient amount of energy during the impact event so as to limit pressure boundary material strains to acceptable levels. Permanent local deformation is expected as a result of accidental drops. However, low strain limits have been established (using a relatively severe SNF and basket load geometry) to reasonably assure that breech of the canister pressure boundary will not occur during impact for a variety of other SNF and internals. Actual drop testing from heights of 9 m (30 feet) with the canister and various internals in multiple essential orientations will be used to validate the containment capability of the DOE SNF canister design. The currently unknown details make it impossible to accurately predict the resulting stress and strain levels in the DOE SNF canister material during impact. Therefore, the design of the specified DOE SNF canister was made robust in order to have a margin between expected strain levels and those strains that could cause a loss of containment in the canister shell.

To support the specified DOE SNF canister design, the following actions should be considered when details and procedures affecting movement of the canister, design of the canister internals, and loading of the canister are finalized:

1. Movements should be planned to minimize the amount of handling of the canister and the number of lifts. Unnecessary lifting should be avoided. Movements should be made at minimum heights, as close to the floor as practical.
2. Movement paths should be defined and restricted to minimize lifts over walls; transport carts; machinery; and other obstructions that would require increased lift heights and impose sharp edges and other puncture threats to the canister.
3. To minimize high lifts over walls, facilities should be designed using a labyrinth arrangement for divider walls when shielding, or other reasons for wall separation are also required. Facilities and support equipment should be designed to eliminate sharp edges, corners, and protrusions. Tops of walls over which lifts may occur should be rounded and covered with replaceable energy-absorbing impact structures.
4. All features of the related facility and support equipment design should give priority to protection of the canister and the SNF.

During accidental drop events, high stress and strain values are expected in local regions of the DOE SNF canister where internal structures contact the canister shell pressure boundary. Internal structures (baskets, spacers, sleeves, dividers, cans, etc.) and fuel elements may contain sharp, stiff elements that pose a puncture threat to the canister containment. Results of the canister drop test evaluations will be contingent on the internal structure designs. Canister internal structure designs must consider fuel safety, condition, stability, and criticality concerns for numerous unique fuel types. A generous volume has been allotted in the DOE SNF canister design to allow flexibility in meeting these demands. Canister protection shall also be considered in the internal structure design requirements. The DOE SNF-specified canister design includes two-inch-thick internal end impact plates (part of the internals design) to protect the dished heads from internal impacts and punctures. Internal baskets, sleeves, dividers, cans, and spacers shall also be designed to: 1) avoid sharp corners and protrusions, and 2) provide large bearing surfaces and energy-absorption features that minimize point loads on the containment shell pressure boundary and resulting high localized regions of stress and strain.

Based on discussions with the repository personnel, once the DOE SNF canister is placed inside the waste package, there are no long-term performance characteristics expected of the DOE SNF canister itself. However, the waste package must satisfy certain design requirements for drop and tip-over accidents. A dropped or damaged waste package may no longer have the required long-term performance capability and may need to be replaced. In order to utilize a new waste package, the SNF must be removed from the damaged waste package. For commercial fuels, this means that the baskets inside of a damaged waste package must limit deformations to such an extent that the commercial fuel can still be removed.

The consequences of a waste package drop or tip-over, however, are not expected to be as severe to the DOE SNF canister as they may be to commercial fuels. If a waste package is dropped or if it tips over, there is still the repository requirement to unload that waste package. Unloading the DOE SNF out of a damaged waste package does not require maintaining such a precise geometry as demanded for commercial fuels. All that is needed is to remove the DOE SNF canister (as well as the HLW canisters) so they can be loaded into a new waste package. That means that the ease of removing the DOE SNF canister depends only on the removal of at least one of the HLW canisters. That one HLW canister could be the canister that was located on top of all the other canisters during the drop or tip-over accident. Being on top, it should be relatively undamaged. Once one HLW canister has been removed, there should be adequate room for removal of the remaining HLW canisters and the DOE SNF canister.

Additionally, the thick, internal end impact plates incorporated into all of the DOE SNF canisters will provide support and resistance to crushing forces generated during the waste package drop or tip-over events. Other canister internal components (i.e., baskets, spacers, sleeves, cans, and dividers) can also be designed to protect the integrity of the canister as well as the fuel. This will help mitigate the consequences of these accident events.

Therefore, the waste package drop or tip-over accident is not considered to be a design-controlling event for the DOE SNF canister itself.

All design concerns resulting from the materials (SNF and internals) placed inside the DOE SNF canisters must be addressed during normal and off-normal conditions of handling, interim storage, transportation, and disposal. The materials to be placed inside of these sealed canisters must be controlled to the extent that adverse or excessive heat, internal pressures, corrosive conditions, and excessive radiation fields are not generated. Because a significant amount of the DOE SNF is highly-enriched uranium (HEU) with ^{235}U enrichment levels greater than 20%, criticality concerns must also be carefully addressed when determining what SNF, how much SNF, and where the SNF is to be placed inside of these canisters. Maintaining SNF placement may be crucial at times for certain fuels, requiring internal devices such as baskets and spacers to not deform under any specified canister loadings.

The design of the SNF, any internals, and the canister itself must be evaluated together. However, the design of the canisters is preceding the design of the canister internals. Internals required for the proper placement of SNF will have to be designed to conform to the design constraints imposed by the existing DOE SNF canisters. This includes the consideration of geometry as bounded by the inner diameter of the canisters; baskets; spacers, or other internal deformations; canister deformations; and canister stresses from SNF or internals loads. Using the above approach, a lightweight, spacious, and robust canister can be designed for DOE SNF, meeting the accident condition demands and satisfying the performance requirements for handling, storage, transportation, and disposal at the repository.

Regulatory guidance

There is little explicit regulatory guidance for design criteria for the standardized DOE SNF canisters. However, 10 CFR Part 72 (USNRC 2003c) regulations describe the design criteria necessary for an interim storage facility to be approved and licensed by the USNRC. The DOE SNF canisters will be part of the components comprising an interim storage facility. However, it is not intended for the DOE SNF canisters to be an interim storage facility by themselves outside a hot cell facility. Furthermore, criteria necessary for transportation casks are provided in 10 CFR Part 71 (USNRC 2003b). By themselves, the DOE SNF canisters are not transportation casks, but will be placed into USNRC-licensed transportation casks. However, 10 CFR Part 71.63(b) (USNRC 2003b) does include double containment for certain SNF. The DOE SNF canisters will not be placed directly into the repository, but will be placed into a disposal container (or waste package after the final barrier weld is accepted) and before being officially disposed. There are generic requirements, such as: confinement or containment barriers for the SNF; material interactions and compatibilities; and the need for criticality control in the regulations. However, there are no complete and explicit design parameters for the DOE SNF canisters.

Design basis

With the assumption that the transportation cask does not provide double containment, the direct placement of damaged SNF into the DOE SNF canister imposes the requirement of using ASME (2001) Boiler and Pressure Vessel Code, Section III design criteria. Certain changes to Section III, Division 3 need to be made in order for the DOE SNF canisters to be fabricated as N-stamped vessels:

1. Division 3 must be changed to allow field operations in order to complete construction (e.g., perform the final closure weld) at locations other than the Code shop.
2. Division 3 must also be changed to allow the actual N-stamping prior to the SNF being loaded into the DOE SNF canisters. In addition, clarifications that allow the use of ultrasonic examination for the final closure weld must be made.
3. Due to the presence of the SNF after loading, the DOE SNF canisters shall not be required to satisfy pressure test requirements of Section III, Division 3, Subsection WB after loading the SNF and final closure welding. Changes permitting helium leak testing, in lieu of pressure testing for low design pressure vessels, must be made to Division 3 rules.

The DOE SNF canisters shall not contain any pressure relief devices. The DOE SNF canisters shall be considered N-stamped (with the Data

Report signed) after the authorized nuclear inspector has accepted the final closure weld. The DOE SNF canisters shall be designed and evaluated according to the acceptance criteria of ASME (2001) Section III, Division 3, Subsections WA and WB. Once storage and transportation details become available, the DOE SNF canisters shall also have the Design Reports incorporate storage and transportation system specifics so that the canisters can be utilized in those systems. The Design Report justifies how the DOE SNF canister can be used. As such, the design loads defined in 10 CFR Parts 71.71 and 71.73 (USNRC 2003b) shall be applied to the DOE SNF canister and the outer transportation cask as a combined system (canister and cask), not individually. The same combined system design philosophy applies to the DOE SNF canister and the applicable storage system. Depending on how the DOE SNF canister is used for interim storage, there could exist the possible requirement to demonstrate DOE SNF canister structural adequacy. If this occurs, it may be necessary to demonstrate compliance with the requirements of ASME (2001) Section III, Division 1, Subsection NB. However, this is achievable because the DOE SNF canisters are already Section III vessels.

Canister design limitations

Reasonable design criteria must be established to adequately address anticipated loadings and events. However, it is unreasonable to incorporate all potential site-specific or facility-specific circumstances into the design bases of these DOE SNF canisters. Hence, it becomes incumbent on all users of these canisters to understand their design limitations and be able to adjust their operations as necessary. One obvious example may be the potential internal pressure. If a specific user intends to load DOE SNF in such a manner that the internal pressure exceeds the maximum considered design pressure, then adjustments must be made to limit the amount of SNF; the amount of included water; or whatever variable that can be adjusted such that the design pressure is not exceeded.

Another more subtle example of user awareness may be the potential for canister puncture during a possible drop event. From a design perspective, geometries of an unlimited number of potential targets that might puncture the canisters are not identifiable. Therefore, the DOE SNF canisters have not been explicitly designed for any specific puncture resistance, although the drop-resistant robust design instills a significant puncture resistance capability by itself. Hence, it will be up to the user to determine if any potential exists for canister puncture during any canister handling operations. If the potential for puncture exists, the user can either perform an analytical evaluation to determine if puncture is possible, or the user can pad the identified target to prevent puncture.

Establishing the acceptability of future use of canisters that may have been subjected to significant loadings imposed while empty or while filled with SNF (but before the top head is attached) becomes the responsibility of the user. Any significant loadings imposed on empty canisters (such as accidental drops) shall be evaluated as described in the final acceptance criteria specified by ASME (2001) Section III, Division 3, Subsections WA and WB (or possibly Section III, Division 1, Subsection NB if appropriate). Significant loadings inadvertently imposed on filled canisters (before the top head is attached) should be evaluated in the same fashion. If a filled canister (before the top head is attached) is "adversely" loaded or dropped, the facility must address the potential cleanup situation under its own operating requirements. Once the canister can be emptied and decontaminated (if necessary), then the user can perform the evaluations and inspections necessary to demonstrate the acceptability of future use or proper disposal of the nonconforming canister.

Although the criteria identified in the preliminary design specification will incorporate reasonable and appropriate design bases, the user must recognize that local conditions and situations must be addressed if they are expected to exceed the specified design criteria. Hence, as indicated before, it becomes incumbent on all users of these canisters to understand the design limitations of the canisters and be able to adjust their operations, where necessary, in order to achieve the expected functionality of these DOE SNF canisters. This includes the proper storage (e.g., dry storage without contamination by damaging environments) of empty DOE SNF canisters prior to loading.

Geometry and materials

All of the DOE SNF canisters shall be right circular cylinders that are able to stand vertically when placed on a flat surface after being loaded

54

with SNF. The large canister shall have a nominal OD of 610 mm (24.00 inches) and a nominal wall thickness of 12.7 mm (0.500 inches). The small canister shall have a nominal OD of 457 mm (18.00 inches) and a nominal wall thickness of 9.53 mm (0.375 inches). Both of these canisters shall be designed for a maximum overall length of either 3,000 mm (118.11 inches) or 4,570 mm (179.92 inches). Dimensional tolerances and fabrication processes (including weld grinding where necessary) shall be controlled so that the maximum dimensions are not exceeded.

Because the material specification is SA-312 (for seamless and welded austenitic stainless steel pipes), certain dimensional tolerances (from material specification SA-530 in Section II of the ASME [2001] Code) associated with the exterior canister shell are already specified. Outer diameter variations shall not be over by more than 3.2 mm (0.125 inches) for the 610 mm (24.00 inches) canister, and 2.4 mm (0.093 inches) for the 457 mm (18.00 inches) canister. For ovality, the difference in extreme outside diameter readings in any one cross-section shall not exceed 1.5% of the specified nominal outside diameter. Straightness tolerances for each canister shall not exceed 3.2 mm (0.125 inches) maximum deviation for every 3.0 m (10 feet) of pipe length (both ends of a 3.0-m straightedge used for the measurement are in contact with the surface). In addition, provisions must be made for the as-welded condition of the final canister weld (attaching the top head to the canister shell). Therefore, an additional diameter increase of 4.76 mm (0.1875 inches) shall be considered acceptable for the crown of the final canister weld. In order to ensure that certain contents can be placed inside these canisters and that the canister can be placed in the repository waste package, the dimensions listed in Table 2 shall be incorporated into the canister design.

Table 2. DOE SNF canister dimensions (DOE 1999).

Canister Size	Nominal Outer Diameter	Long Canister Max. External Length	Short Canister Max. External Length	Min. Internal Diameter	Long Canister Min. Internal Length	Short Canister Min. Internal Length
Large	610 mm (24.00 in.)	4,570 mm (179.92 in.)	3,000 mm (118.11 in.)	579 mm (22.80 in.)	4,038.6 mm (159 in.)	2,470.2 mm (97.25 in.)
Small	457 mm (18.00 in.)	4,570 mm (179.92 in.)	3,000 mm (118.11 in.)	430 mm (16.93 in.)	4,114.8 mm (162 in.)	2,540.0 mm (100 in.)

Current repository criteria indicate that low-carbon austenitic stainless steel or stabilized austenitic stainless steel materials are acceptable. The National Spent Nuclear Fuel Program (NSNFP) is proceeding on the assumption that low-carbon austenitic stainless steel will be acceptable to the repository. Although canister specific drop tests have been performed that clearly demonstrate the acceptability of 304L, stainless steel 316L has better resistance to pitting corrosion and hydrogen embrittlement. Considering the variations in the actual material properties of stainless steel, it is believed that the use of 316L would be an acceptable canister material, even considering accidental drop events. Therefore, the DOE SNF canisters shall be made of SA-312, type 316L stainless steel for the shell; and SA-240, type 316L for all other parts, including the heads, name-plates, and lifting rings. The optional plugs and plug thread plates shall be SA-479, type 316L stainless steel. All stainless steel materials shall be annealed and pickled. The DOE SNF and canister internals shall preclude chemical, electrochemical, or other reactions (such as internal corrosion) of the canister or waste package such that there will be no adverse effects.

As part of the material selection process, important parameters to consider are the anticipated erosion and corrosion values expected during interim storage and transportation use. A total value of 1.27 mm (0.050 inches) of pressure boundary wall thickness reduction has been established as the erosion and corrosion value to be used for canister design purposes. This corrosion/erosion value reflects the full design lifetime of 100 years. Therefore, prior to acceptance at the repository, the DOE SNF canisters shall be protected from adverse environmental conditions in such a manner as to prevent the total wall thickness corrosion/erosion limit from being exceeded. A 50-year interim storage and transportation interval shall be assumed for this specific wall thickness reduction evaluation. The assumption is made that once the DOE SNF canister is placed inside the waste package, insignificant corrosion or erosion will occur for the next 50-year interval.

The canisters will be subjected to a radiation environment. Because criticality is to be eliminated, the canisters should not be exposed to large neutron fluences. Long-term cumulative exposures to high neutron fluence (on the order of 10^{17} n/cm^2 and greater) has caused degradation in reactor vessels, but this is not applicable to the DOE SNF canisters.

Radiation fields (10^7 rad/h or less) are expected, but no significant material damage or degradation is anticipated for the stainless steel material.

Both the inner and outer surfaces of the canisters shall have a finished condition such that acceptable nondestructive examinations can be performed in order to satisfy ASME (2001) Code, Section III, Division 3, Subsection WB requirements. However, no specific surface finish is specified for the DOE SNF canisters. The repository has limits on outside surface contamination for the canister, but has not imposed an associated surface finish in conjunction with the surface contamination requirement. The repository does require that any burrs, sharp edges, and weld edges not exceed 0.5 mm (0.0197 inches). The interior surfaces shall be smooth enough to allow easy loading of any DOE SNF or internals (baskets, spacers, sleeves, dividers, cans, etc.) so as to not damage the SNF.

Contents

The contents of the canister—including SNF and any applicable internals necessary for the safe placement and orientation of the DOE SNF—have not been specifically defined. However, it is assumed that the contents will not compromise the structural integrity of the DOE SNF canister. It is also assumed that the contents will satisfy all applicable regulations and requirements, especially those set forth by the repository. For example, the contents of the disposal canister shall contain no pyrophoric, combustible, explosive, or chemically-reactive materials in an amount that could compromise surface-facility or repository preclosure safety, or repository long-term performance. The DOE SNF canisters shall be designed for the total maximum allowable weights (canister plus contents) listed in Table 3. These weight limits are equal to or less than the weight limits established in the repository's Interface Control Document (ICD).

Table 3. DOE SNF canister maximum total allowable weights (DOE 1999).

Canister Size	Nominal Outer Diameter	Long Canister Maximum Total Weight	Short Canister Maximum Total Weight
Large	610 mm (24.00 in.)	4,535 kg (10,000 lb$_f$)	4,080 kg (8,996 lb$_f$)
Small	457 mm (18.00 in.)	2,721 kg (6,000 lb$_f$)	2,270 kg (5,005 lb$_f$)

When loading the DOE SNF, the center-of-gravity of the entire contents (SNF, baskets, spacers, sleeves, dividers, etc.) shall be within 127.0 mm (5 inches) of the canister centerline for the 457 mm (18.00 inches) nominal OD canister, and within 203.2 mm (8 inches) of the canister centerline for the 610 mm (24.00 inches) nominal OD canister. These loading restrictions are to avoid excessive lopsided loading situations and to limit resulting stresses in the lifting ring and the adjacent skirt portion of the canister. When possible, the center-of-gravity of the loaded DOE SNF will be as close to the canister centerline as reasonably achievable. The axial location of the center-of-gravity of a loaded DOE SNF canister shall be within 609.6 mm (24.0 inches) of the canister centroid. Sites performing SNF loading may make separate evaluations of center-of-gravity locations if the indicated center-of-gravity limitations are exceeded.

Canister sealing

Storage, transportation, and repository disposal criteria all indicate specific requirements associated with the safe and proper sealing of SNF containers. After loading the SNF, the DOE SNF canisters shall be covered with an inert gas (e.g., helium) inside of the canister. Such a gas is intended to eliminate or significantly reduce SNF corrosion; provide more appropriate heat transfer conditions internally; and reduce combustion concerns.

Sealing of the DOE SNF canisters may be required for interim storage, depending on the type of storage system utilized. However, if the storage system permits it, the DOE SNF canisters may be sealed after interim storage. Incorporated into the DOE SNF canister design is the option of a threaded plug in the top and bottom head. However, prior to transportation and repository disposal, the DOE SNF canister must be seal-welded closed. The DOE SNF canisters shall be backfilled with an inert cover gas (e.g., helium) inside of the canister at a pressure of 13.8 to 27.6 kPa (2 to 4 psig). The final canister weld (attaching the top head to the canister shell) shall implement a welding procedure that can be qualified to yield leak-tight welds. At a minimum, in order to demonstrate compliance with the ASME (2001) Code and obtain the Code stamp per the proposed Code changes, the DOE SNF canister shall be helium leak tested to verify that no leakage is detected that exceeds the rate of 10^{-4} std cm³/s. After closure of the canister, the DOE SNF canister shall not contain or generate free gases other than air, inert cover gas, and radiogenic gases with an immediate internal gas pressure not to exceed 151.7 kPa (22 psig). Therefore, the DOE SNF canister shall have a maximum allowable (design) pressure of 344.8 kPa (50 psig).

Incorporated into the DOE SNF canister design is the option of a threaded plug in the top and bottom head. These threaded plugs can be used, where necessary, for a number of functions such as: canister draining, degassing, and remote inspection. Because the containment feature of the canister depends upon the proper installation of the threaded plug, installation or removal of the threaded plug(s) is expected to be performed while the DOE SNF canister is inside a hot cell. When using these threaded plugs, it is necessary to seal-weld the threaded plugs in order to establish an acceptable containment boundary per ASME (2001) Code, Section III in requirements. The seal weld shall cover any exposed threads on the plug.

Shielding

For the purposes of the design specification, it is assumed that the DOE SNF canisters do not require additional shielding beyond that provided by the canisters themselves, or in conjunction with shielding provided by the facilities handling the DOE SNF canisters, the storage system, or transportation system. It is assumed that any of the facilities handling the DOE SNF canisters or the repository will have adequate equipment necessary to handle the DOE SNF canisters by themselves. If additional shielding is required by local sites or facilities, it is up to the user to provide adequate shielding measures without adversely affecting the DOE SNF canisters.

Criticality

For the purposes of the design specification, it is assumed that adequate attention to the types and amounts (proper fissile limits) of SNF to be loaded into the canisters (or proper configuration) using properly-designed internals (baskets, spacers, sleeves, dividers, cans, etc.) will preclude any criticality concerns. Therefore, this design specification assumes no criticality events during the canister's design life. For criticality concerns, the DOE SNF canisters must only be capable of maintaining reasonable geometric integrity. The personnel responsible for any designs of internals necessary for criticality prevention must address any associated concerns should the DOE SNF canisters be loaded, dropped, or handled in such a fashion as to adversely load the SNF or internals.

Normal operating loads and environmental conditions

The weight considerations listed in Table 3 are for all of the DOE SNF canister geometries. The DOE canisters shall not be horizontally or vertically stacked at any time with any other canisters without a proper evaluation of all possible consequences. Interim storage, transportation, and disposal scenarios are situations where the canisters are within other enclosures or facilities, and these canister placements shall be properly evaluated. The limitation of no vertical or horizontal stacking is imposed not from a strength concern, but rather from a safety viewpoint. Vertically stacking these long, slender canisters (either empty or filled) would prove to be a major safety concern for personnel due to a lack of stability. Horizontally stacking these canisters (like a cord of firewood) could be permissible in an empty condition, but that could also be unstable due to rolling concerns unless specific actions are taken to prevent it from happening. Horizontally stacking these canisters, when filled with SNF, appears to be undesirable from a rolling stability concern. Potential criticality implications and excessive heat generation concerns may exist if the canisters were to be stacked in close proximity to each other. Therefore, due to a number of unknown implications and concerns, the preliminary design specification assumes that the canisters will not be vertically or horizontally stacked. If the canister user needs to stack the DOE SNF canisters, then the user must provide the justification, addressing all potential implications.

The lifting fixture is not part of the canister design specification. However, the design of the lifting fixture does have an effect on the resulting design of the canister. The lifting fixture shall provide at least three locations to engage the canister lifting ring, with the capability to engage and disengage remotely. The lifting fixture shall be capable of engaging and disengaging while remaining within the projected perimeter of the DOE SNF canisters.

The DOE SNF canisters shall be designed to be vertically lifted with a lifting fixture that engages underneath the 12.7-mm (1/2-inch) thick lifting ring. Material temperature limits for lifting the canisters shall be 148.9°C (300°F). Due to the symmetry of the specified DOE SNF canister design, either end shall be capable of being used to lift the canister. The Maximum Normal In-Plant Handling Pressure (MNIP) shall be considered as acting coincidentally. With respect to recovering from an accidental canister drop or tip-over (regardless of severity), the canisters shall be designed to be picked up from both extreme ends or tilted back upright from a horizontal position. Stresses resulting from this action shall satisfy normal operating condition stress limits defined in ASME (2001) Code, Section III, Division 3, Subsection WB. Worst case temperatures and pressures shall be considered as acting coincidentally. The weight of the contents shall be assumed to be lumped at the centroid of the canister.

The MNIP is the maximum pressure that would develop in a DOE SNF canister during initial handling, interim storage, transportation; initial repository handling; or disposal container loading prior to actual emplacement in a repository drift under the most severe conditions of normal in-plant handling operations. The DOE SNF canister shall be designed for a MNIP not to exceed 344.8 kPa (50 psig) per the criteria of the ASME (2001) Code, Section III, Division 3, Subsection WB.

The Maximum Normal Operating Pressure (MNOP) is the maximum pressure that would develop in a DOE SNF canister during initial handling, interim storage, transportation, or initial repository handling or disposal container loading prior to actual emplacement in a repository drift without venting. The DOE SNF canister shall be designed for an MNOP not to exceed 151.7 kPa (22 psig) per the criteria of the ASME (2001) Code, Section III, Division 3, Subsection WB.

The Primary Service Temperature for a DOE SNF canister, when it is not inside any other container, is 176.7°C (350°F), and 343.3°C (650°F) after placement within another enclosed container (a storage industry canister for interim storage or a transportation cask), possibly with other heat generating DOE SNF canisters or HLW canisters. The maximum operating temperature for a DOE SNF canister when it is not inside any other container, is 148.9°C (300°F), and 315.5°C (600°F) after placement within another enclosed container (a storage industry canister for interim storage or a transportation cask), possibly with other heat generating DOE SNF canisters or HLW canisters.

The DOE SNF canisters shall be designed for 20 full MNIP and temperature cycles of a canister achieving its maximum steady state-operating temperature of 315.5°C (600°F) inside another container, and then suddenly being exposed to an external calm air temperature environment of 10°C (50°F) while the canister simultaneously loses its internal pressure. The maximum thermal gradient associated with this event shall be evaluated per the criteria of the ASME (2001) Code, Section III, Division 3, Subsection WB. If the canisters are subjected to any other significant fatigue loads due to initial SNF loading, interim storage, transportation, or loading into a disposal container at the repository, a detailed fatigue analysis shall be performed per WB-3221.9(e). If necessary, cumulative usage factors from all uses (SNF loading, canister handling, storage, or transportation) shall be evaluated once these values are known.

The DOE SNF canisters shall be capable of maintaining containment in temperature environments that range from –40°C to 343.3°C (–40°F to 650°F), excluding accidental drop scenarios or other accidental events when being handled by itself or inside of the waste package at the repository.

For situations requiring specific design evaluations where significant compressive stresses occur in the canister, the buckling stress shall be taken into account. Buckling situations need to be considered in terms of being able to remove the canister from the enclosing container (either the storage industry canister or the transportation cask).

The DOE SNF canisters shall be designed for any other normal operating condition loads resulting from initial handling, interim storage, transportation, or handling and loading into a disposal container at the repository once the loads and environments have been defined.

Hypothetical accident loads and environmental conditions

Anytime a loaded DOE SNF canister is being handled by itself, the canister shall be within a HEPA-filtered building or facility. This eliminates the requirement of specifically designing the DOE SNF canisters to any specific ASME (2001) Code stress limits for accidental drop events. However, when the DOE SNF canister is enclosed within a storage industry canister for interim storage purposes, or within a transportation cask, the DOE SNF canister shall be designed in accordance with the criteria in the ASME (2001) Code, Section III, Division 3 stress limits identified in WB-3224 or Section III, Division 1, Subsection NB-3225 as required. For the repository waste package drop or tip-over event, the canister shall be considered adequate as-developed since the only requirement is to remove the DOE SNF canister from the damaged waste package and place it into another undamaged waste package. The DOE SNF canister has a robust design because it was developed to maintain containment when subjected to a 9-meter (30-foot) drop onto an essentially unyielding surface. It is this robust design that makes the DOE SNF canister adequate to survive the waste package drop or tip-over event. For situations requiring specific design evaluations where significant compressive stresses occur in the canister, the buckling stress shall be taken into account. Buckling situations need to be considered only in terms of being able to remove the canister from the enclosing container (either the storage industry canister or the transportation cask).

Quality assurance

The designer and fabricator of the DOE SNF canisters shall establish, maintain, and execute a quality assurance program based on the criteria necessary to satisfy ASME (2001) Code, Section III, Division 3 construction criteria, 10 CFR Parts 71 (Subpart H) (USNRC 2003b), and 72 (Subpart G) quality assurance requirements.

Physical protection of SNF

Because the DOE SNF canisters are to be seal-welded for transportation and repository disposal, the DOE SNF canisters do not require the use of an USNRC approved tamper-safe seal. Tamper-indicating devices are only required on canisters containing strategic special nuclear material that are not seal-welded. Depending on how the DOE SNF canister is being used within an interim storage system, a tamper-safe seal may be required if the DOE SNF canister is not seal-welded.

Labeling

The DOE SNF canisters shall be capable of being properly labeled as follows:

1. The labels shall be an integral part of the canister, engraved to a depth no greater than 0.8 mm (1/32 of an inch) that can be reasonably expected to remain legible for 100 years at temperatures of 25°C (77°F) to 400°C (752°F)
2. The labels shall have a unique alphanumeric identifier
3. The labels shall not impair the integrity of the canister
4. The labels shall be chemically compatible with the canister material
5. The top label shall be visible from the top of the canister with the lifting fixture engaged, with characters approximately 25.4 mm (1 inch) in height
6. The labels shall not cause the canister dimensional limits to be exceeded

The DOE SNF canisters should be labeled on the outer most surface of each lifting ring on the top and bottom ends of the canister (two places). The alphanumeric identifier shall be readable as if the remotely-operated cameras are on the outside of the canister looking inward toward the axial centerline of the canister. Placement of an alphanumeric label on the lifting ring shall not cause any interference or loading concerns.

Documentation

The designer of the DOE SNF canisters shall provide adequate documentation, reports, and design drawings in the proper form and format to satisfy the proper quality assurance requirements and record storage requirements. In addition, adequate documentation shall also be provided that:

1. Supports the acceptable design of the DOE SNF canisters
2. Permits an independent review of all design procedures and calculations
3. Identifies all software used in the design process

4. Indicates appropriate validation and verification documentation of all software used for the design calculations
5. Indicates the involved design personnel has the required experience, education, training, and proficiency
6. Is legible and in a form suitable for reproduction, filing, and retrieval

THE WELDING, INSPECTION, AND REPAIR SYSTEM FOR STANDARDIZED SPENT NUCLEAR FUEL CANISTERS OF U.S. DEPARTMENT OF ENERGY

INTRODUCTION

The National Spent Nuclear Fuel Program (NSNFP) of the U.S. Department of Energy (DOE) is developing a set of standard canisters for spent nuclear fuel (SNF). Seal welding of the SNF standardized canisters must be performed after DOE SNF is loaded into the containers. As such, the welding and nondestructive examination of the acceptability of the results of the welding will be performed remotely in a hot cell. Because the containers may be part of a storage, transport, and/or disposal system during their useful lifetime, it is DOE's intent to design and build the vessels according to the ASME (2001) Section III, Division 3 requirements. The final seal weld for closure needs to meet these requirements.

Key issues regarding the welding and nondestructive examination of the standardized DOE SNF canisters' final seal welds include the following:

1. Remote operability in a high-radiation environment (11,195 rad/h gamma radiation)
2. Resultant weld quality and documentation
3. Long-term thermal/micro-structural stability of materials
4. Storage environment
5. Final weld joint design of the standardized canister

The plan is directed towards developing welding, nondestructive examination, and analysis techniques (including equipment) that can not only be used concurrently to weld and inspect the final closure seal welds on the standardized DOE SNF canisters, but also addresses the aforementioned key issues. The SNF welding, inspection, and repair system compliance and verification matrix is presented in Table 4.

Table 4. SNF welding and inspection system compliance and verification matrix (Watkins 2003b).

Requirement Number	Requirement	Compliance Method	Test	Demo	Analysis	Inspect
WELDING AND INSPECTON						
NSF1	The closure cell system shall weld the DOE Standardized Canister lid.	The closure cell system will weld the canister closure welds with the Welding and Inspection System (W&IS).		X		
NSF2	All welds will be made using the cold-wire gas tungsten arc welding process.	The W&IS will incorporate cold-wire gas tungsten arc welding equipment and use solid filler wire that meet ASME Section II, Part C.		X		
NSF3	The W&IS shall be capable of welding and repairing defective welds.	The W&IS includes grinding, welding, and inspection capabilities to support weld repair operations. The Welding and Inspection System will be capable of performing repair activities in accordance with requirements of the ASME Boiler and Pressure Vessel Code, Section III, Division 3, Subsection WC, 2001 Edition.		X		
NSF4	The W&IS shall be developed and/ or qualified in accordance with quality requirements, as applicable.	Welding procedures and performance qualifications will be performed in accordance with ASME Boiler and Pressure Vessel Code, Section III, Division 3, Subsection WC, 2001 Edition.		X		X
		Nondestructive inspection methods and performance qualifications will be performed in accordance ASME Boiler and Pressure Vessel Code, Section III, Division 3, Subsection WC, 2001 Edition.				

68

Table 4. (cont'd).

Requirement Number	Requirement	Compliance Method	Test	Demo	Analysis	Inspect
WELDING						
NSF 5	The W&IS shall deposit narrow-groove closure welds.	The W&IS will be designed to weld the Standardized Canister closure weld.		X		
NSF 5.1	The W&IS shall tack weld the lid to Standardized Canister shell in the horizontal position.	See NSF 5		X		
NSF 5.2	The W&IS shall deposit the closure weld on the standardized canister shell in the horizontal position.	See NSF 5		X		
INSPECTION						
NSF 6	The W&IS shall inspect narrow-groove welds.	The system will be designed to inspect the Standardized Canister closure weld.		X		
NSF 6.1	The W&IS shall inspect tack welds of the closure weld on the Standardized Canister in the horizontal position.	The closure tack will be visually inspected a surface profiler.		X		
NSF 6.2	The W&IS shall inspect the closure weld on the Standardized Canister in the horizontal position.	The surface of the multi-pass, full depth, backed, narrow groove closure weld will be visually inspected. The surface of this weld will also be eddy current inspected. The volume of this weld will also be ultrasonically inspected.		X		

69

Table 4. (cont'd).

Requirement Number	Requirement	Compliance Method	Test	Demo	Analysis	Inspect
INSPECTION						
NSF 6.3	The W&IS shall inspect the closure weld on the Standardized Canister in the horizontal position.	The partial volume of the multi-pass, full depth, backed, narrow groove closure weld will be ultrasonically inspected on a pass-by-pass basis. This will be done subsequent to completion of each weld pass. Inspection is limited to a subset of sound paths not obstructed by unwelded portions of weld joint.		X		
NSF 6.4	The Welding and Inspection System shall inspect repair cavities and associated repair welds for nonconformities.	Repair cavities will be inspected by ET. Repair welds will be inspected in the same manner as the original weld.		X		
WELDING AND INSPECTION HARDWARE						
NSF 7	The W&IS shall have adequate functionality to perform the required welding and weld inspections.	The Welding and Inspection System will have capability of: System Controls Motion Control Process Control Inspection Sensors Maintainability.		X		
CONTROLS						
NSF 7.1	The W&IS shall have means of controlling the system.	The operator shall be able to program or operate both the torch movements and the welding parameters from the control console.		X		
NSF 7.1.1	The W&IS shall have capabilities for integration with other closure cell systems.	A standard method (e.g., OPC [OLE for Process Control]) will be used for passing data and commands from one subsystem to another.		X		

70

Table 4. (cont'd).

Requirement Number	Requirement	Compliance Method	Test	Demo	Analysis	Inspect
CABLE MANAGEMENT						
NSF 7.2	The W&IS shall have means of managing the various cables and hoses necessary for operation of the in-cell portions of the system.	The W&IS shall utilize wireless Ethernet and will have system for management of cables and hoses where necessary.		X		
MOTION CONTROL						
NSF 7.3	The W&IS shall have means of providing the motion and motion control necessary for the operation of the in-cell portions of the system.	The W&IS will have a circumferential table approximately located at the top of the WP. The circumferential table has a center hole for the canister to protrude through. Three equipment towers are mounted on the circumferential table. Rotation of the circumferential table provides circular movement around the canister.		X		
NSF 7.3.1	The W&IS shall incorporate sufficient range of motion to be able to position the End Effectors at the proper locations to carry out their functions.	Three equipment columns will be utilized and will provide sufficient range of motions to weld and inspect the closure weld joint.		X		
WELDING HARDWARE						
NSF 7.4	The W&IS shall have the necessary functionality needed for the welding process.	The W&IS System will have Welding End Effectors for welding canister closure welds.		X		

71

Table 4. (cont'd).

Requirement Number	Requirement	Compliance Method	Test	Demo	Analysis	Inspect
WELDING HARDWARE						
NSF 7.4.1	The W&IS shall have the necessary functionality for performing the required welds.	The Welding End Effectors will have the following capabilities: Air cooled torch Remotely adjustable filler wire guide Manual seam tracking Seam tracking/torch oscillation mechanism Weld vision cameras Thermocouple probe Arc voltage control/arc touch starting Quick disconnect mount.		X		
NSF 7.4.1.1	The W&IS shall use an air cooled welding torch.	An air cooled welding torch has been designed for this application.		X		
NSF 7.4.1.2	The W&IS shall incorporate a remotely adjustable filler wire guide.	A commercially available remotely adjustable filler wire guide has been selected for this application.		X		
NSF 7.4.1.3	The W&IS shall incorporate seam tracking/torch oscillation capability.	A manual seam tracking mechanism has been designed for this application.		X		
NSF 7.4.1.4	The W&IS shall incorporate weld vision cameras capable of viewing the weld pool with arc light attenuation.	Commercially available weld vision cameras (front and rear) have been selected for this application.		X		
NSF 7.4.1.5	The W&IS shall incorporate capability of measuring the weld joint temperature prior to welding.	A commercially available thermocouple probe has been selected for this application.		X		

72

Table 4. (cont'd).

Requirement Number	Requirement	Compliance Method	Test	Demo	Analysis	Inspect
WELDING HARDWARE						
NSF 7.4.1.6	The W&IS shall incorporate capability of automatically controlling arc voltage and starting the welding arc by means of the touch starting method.	An electro-mechanical mechanism has been designed for this application.		X		
NSF 7.4.1.7	The W&IS shall incorporate a means of disconnection from the equipment towers.	A commercially available quick disconnect has been selected for this application.		X		
NSF 7.4.2	The W&IS shall incorporate cold wire, direct current, electrode negative (DCEN) GTAW power supplies having adequate functionality to perform the needed welding.	The welding power supplies will have a rated output of 450 A at 25 V @ 100% duty cycle (minimum) They will have capability of complete control via an analog or Ethernet interface (T-base 100 or 1000).		X		
NSF 7.4.3	The W&IS shall incorporate filler wire feeders having adequate functionality.	The filler wire feeders will have remote control capability, handle wire of 0.035 to 0.045 in. diameter, and feed wire at speeds from 25 to 300 ipm.		X		
WELD INSPECTION HARDWARE						
NSF 7.5	The W&IS shall have the necessary functionality needed to perform the required weld inspections.	The Inspection End Effectors have the following capabilities: VT inspection end effector for visual inspection, ET inspection end effector for eddy current inspection of repair cavities surfaces in closure weld, and an UT/ET end effector for ultrasonic and eddy current inspection of the final surface of the closure weld.		X		

73

Table 4. (cont'd).

Requirement Number	Requirement	Compliance Method	Test	Demo	Analysis	Inspect
WELD INSPECTION HARDWARE						
7.5.1	The VT inspection end effector shall have functionality to inspect welds visually.	A commercial available laser profiling and sensor on the welding end effector shall be used for visual inspection.		X		
7.5.2	The ET inspection end effector shall have functionality to inspect the completed surface of the closure weld and repair cavities.	The ET inspection end effector shall have the following capabilities: • Repair groove inspection probe • Camera for monitoring • Manual oversight of tracking • Suspension for compliance in probe placement.		X		
7.5.2.1	The ET inspection end effector shall incorporate a repair groove probe with the functionality to inspect standard repair groove for closure weld.	Probe will be designed to comply with inspection requirements. An engineering design file will document the approach.		X		
7.5.2.2	The ET inspection end effector shall incorporate a camera for viewing and manual tracking of the travel of the end effector.	A commercially available radiation hardened camera has been selected for this application.		X		
7.5.2.3	The ET inspection end effector shall incorporate functionality to insure probe remains in contact with surface being inspected.	Suspension designed to ensure consistent contact is maintained.		X		

Table 4. (cont'd).

Requirement Number	Requirement	Compliance Method	Test	Demo	Analysis	Inspect
WELD INSPECTION HARDWARE						
7.5.3	The UT/ET inspection end effector shall have functionality to inspect completed closure narrow groove weld with eddy current and ultrasonic inspection and a partially completed closure groove weld with ultrasonic inspection.	The UT/ET inspection end effector will have the following capabilities: • UT probes for inspection of completed and partially completed closure weld • ET probe for eddy current inspection of completed closure weld.		X		X
7.5.3.1	The UT/ET inspection end effector shall have UT probes with functionality to inspect the completed and partially completed closure weld.	Probes will be designed to comply with ASME Section III, Division 3, Subsection WB inspection requirements. An engineering design file will document the approach.		X		X
7.5.3.2	The UT/ET inspection end effector shall have ET probes with functionality to inspect the surface of the completed outer lid weld.	Probes will be designed to comply with ASME Section III, Division 3, Subsection WB inspection requirements. An engineering design file will document the approach.		X		X
7.5.3.3	The UT/ET inspection end effector shall incorporate a camera for viewing and manual tracking of the travel of the end effector.	A commercially available radiation hardened camera has been selected for this application.		X		
7.5.3.4	The UT/ET inspection end effector shall incorporate functionality to insure probes remain in contact with surface being inspected.	Suspension designed to ensure consistent contact is maintained.		X		

75

Table 4. (cont'd).

Requirement Number	Requirement	Compliance Method	Test	Demo	Analysis	Inspect
REMOTE MAINTENANCE						
NSF 7.6	The W&IS shall have the necessary functionality needed for maintenance.	The W&IS will be manually recoverable and maintainable in either the closure cell maintenance area or the glovebox.		X		
NSF 7.6.1	All welding equipment located in the closure cell will be designed for manual recovery.	The various end effectors may be recovered manually.		X		
NSF 7.6.2	Welding and Inspection subsystems used in the closure cell requiring frequent maintenance or servicing will be designed for glove box maintenance.	The welding and inspection end effectors will be designed for glove box maintenance.		X		
NSF 7.6.3	W&IS components not designed for glovebox maintenance or servicing will be designed for remote or in-cell maintenance.			X		
RAD ISSUES						
NSF 8	The W&IS System shall be designed to minimize waste streams during operational activities.			X		
NSF 8.1	For decontamination reasons, tooling and equipment surface finishes shall be 32 micro-inches or better.			X		

76

Table 4. (cont'd).

Requirement Number	Requirement	Compliance Method	Test	Demo	Analysis	Inspect
RAD ISSUES						
NSF 8.2	Tooling and equipment materials shall have hard surfaces to simplify decontamination,	Only components whose function may be affected by surface hardening will be used in a surface hardened version in the system.		X		
NSF 8.3	Equipment and tool designs shall minimize the potential to trap radioactive particles.			X		
NSF 8.4	Use of materials that degrade in high radiation fields, such as polymers and many lubricants, shall not be used unless a suitable substitute is not available.			X		
SAFETY ISSUES						
NSF 9	The W&IS shall be designed for safe operation in accordance with Safety in Welding, Cutting, and Allied Processes, ANSI Z49.1, latest edition, where applicable.	The W&IS uses industry accepted, commercially available welding hardware to ensure safe operation.				X
NSF 9.1	The W&IS shall be designed for safe operation in accordance with laser safety requirements.	The W&IS uses industry accepted, commercially available surface profiling hardware to ensure safe operation.				X

77

Table 4. (cont'd).

Requirement Number	Requirement	Compliance Method	Test	Demo	Analysis	Inspect
GENERAL DESIGN						
NSF 10	The W&IS shall be designed for an operating life of 30 years. This design requirement affects facility structures, utilities, and major equipment. Equipment items to be specified with an operating life of less than 30 years shall be identified on a spare parts list during the development of equipment specifications.	The W&IS will be designed in a modular fashion that will support continual upgrading of the technology over the 30-year operating life.		X		
NSF 10.1	The W&IS shall be capable of performing welding activities in the closure cell environment.	The W&IS is designed to weld in the closure cell.		X		
NSF 10.2	All welding subsystems will be capable of meeting productivity and quality requirements in the closure cell.	The welding system will be designed for the application using industry-accepted technologies that should meet the productivity and quality requirements.		X		
NSF 10.3	The W&IS shall be capable of cleaning the weld joint and adjacent areas by wire brushing.	A weld dressing end effector will be included in each welding system. These end effectors will include wire-brushing capabilities.		X		
NSF 10.4	Materials, tools, and equipment in the closure cell shall not pose a risk of contaminating the WP with foreign elements that could promote corrosion or poor welds (e.g., sulfur compounds, hydrocarbons, zinc, and halides.)			X		

Table 4. (cont'd).

Requirement Number	Requirement	Compliance Method	Test	Demo	Analysis	Inspect
GENERAL DESIGN						
NSF 10.5	Tool contact on and near the weld preps prior to welding shall be avoided to protect the weld zone from damage and impurities.	With the exception of the temperature measuring transducers, tool contact on or near the weld preps prior to welding will be minimized.		X		
NSF 10.6	The W&IS will be designed to prevent unintentional release of liquids into the closure cell.			X		
NSF 10.6.1	A small TBD amount of water may be used as a couplant in the ultrasonic inspection probes.	Maximum of TBD (ml) of water will be introduced to the cell as a couplant through a probe membrane.		X		
DOCUMENTATION						
NSF 11	The W&IS shall be capable of acquiring data, as applicable, to support WP final documentation.	The W&IS will be capable of determining or receiving coordinate information necessary to meet this requirement.		X		

79

As shown in Figs. 3–6, the weld inspection equipment is stationed on a carrousel that rotates around the SNF standardized canister. The weld power supply and inspection electronics are housed behind a shielding device to minimize the equipment exposure to a high-radiation environment. Each one of the three towers of the carrousel supports one component (i.e., welding, inspection, repair) of the system. Only the end effector portion of the equipment is exposed to a high radiation environment. All repair cavities will have a consistent geometry. The Eddy Current Testing (ET) probe will be shaped to fit the groove geometry. Iterative grind and inspect steps will be performed to verify the removal of a defect.

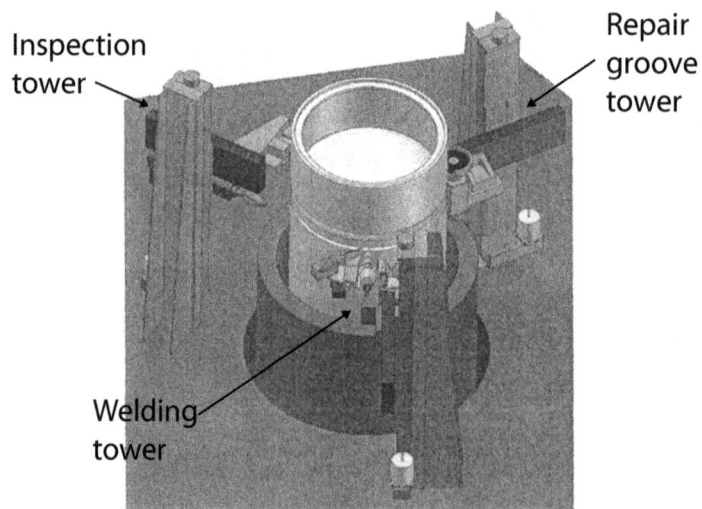

Fig. 3. The welding, inspection, and repair system (McJunkin 2003).

80

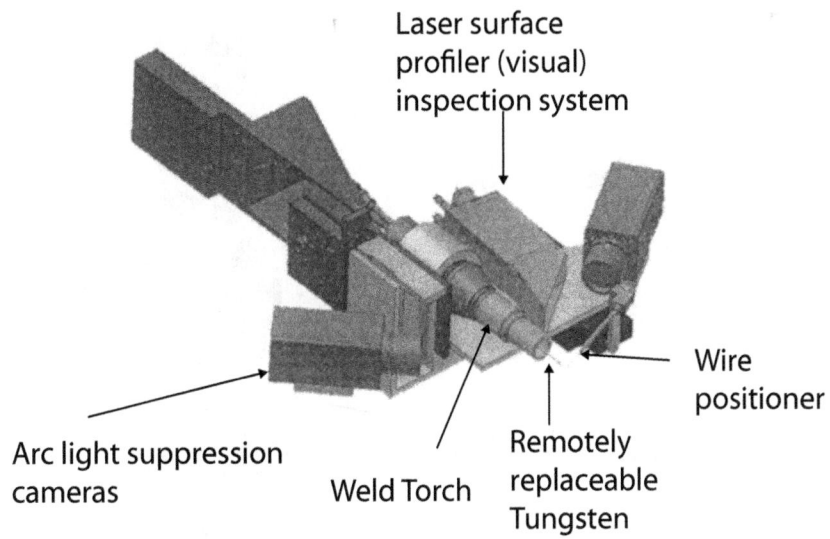

Fig. 4. The welding subsystem (McJunkin 2003).

Repair groove grinder

Fig. 5. The repair subsystem (McJunkin 2003).

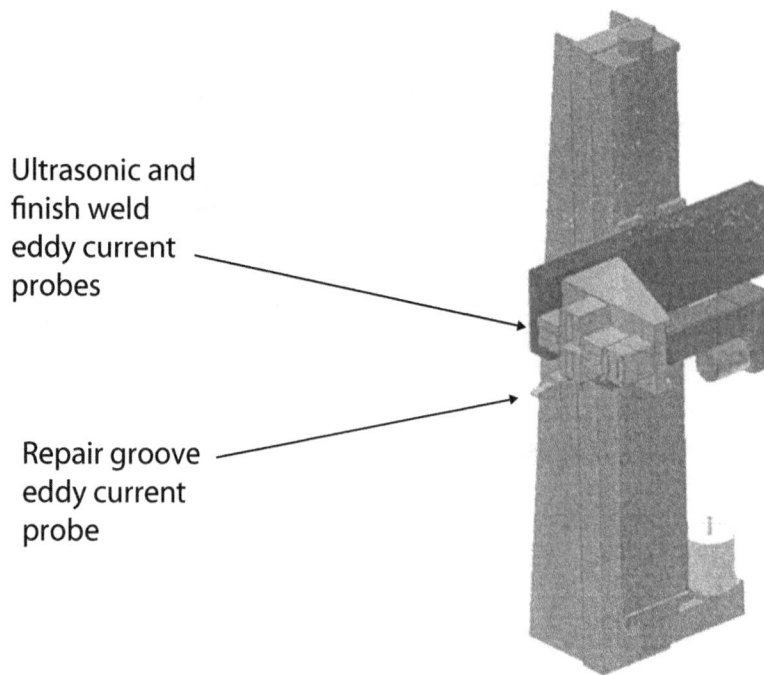

Fig. 6. The inspection subsystem (McJunkin 2003).

ASSESSMENT OF AVAILABLE COMPETENCY

One of the key issues identified during the initial phases of the project was the need to evaluate the availability of relevant commercial competency for various aspects of the project. Subsequently, three areas were identified for assessment as follows:

1. Components of the welding
2. Inspection
3. Repair system

In the next step, requirements for each area were defined. Note that the requirements listed for each component include general requirements as well as environmental requirements. The environmental requirements are based on assumptions made on the fuel to be contained and the expected temperature. Finally, the Project Team performed vendor assessments for each area. In the following the technical criteria are described to be followed by the capabilities of each vendor.

Welding power supplies

Technical requirements: The identified technical requirements are as follows:

1. Radiation environment: N/A (additional shielding in place to limit dose)
2. Rated Output: 450 A/25 V at a minimum of 100% duty cycle
3. Remote contractor control
4. Remote control of current
5. Pulse current capability
6. Input voltage: 460 V, 3-Phase capable
7. An analog and/or Ethernet interface T-based 100 or 1000
8. Welding power supply must be able to support cold wire gas tungsten arc welding (GTAW) and direct current electrode negative (DCEN)

Vendor 1: Lincoln Electric "Powerwave 455M":

1. Domestic supplier
2. Meets above criteria
3. Has numerous interfaces (e.g., Ethernet, Device Net, and serial port)
4. Can also be controlled via analog and digital signals
5. Solid state
6. Has electrical wave form control
7. Has Lab Windows libraries

Vendor 2: Miller Electric "XMT 456 CC/CV" and "Dimensions 452":

1. Domestic supplier
2. Meets above criteria
3. Can also be controlled via analog and digital signals
4. Has electrical wave form control
5. XMT 456 CC/CV is an inverter power supply; Dimensions 452 is solid state

Vendor 3: Thermal Dynamics "Power Master 500" and "Power Master 500 P":

1. Domestic supplier
2. Both of these power supplies meet criteria
3. Both have electrical wave form control
4. Can also be controlled via analog and digital signals
5. Both are inverters
6. PM 500P has pulser in the power supply, while PM 500 does not

Conclusion: The welding power supplies that are commercially available meet the DOE requirements for this application.

Welding torches

Technical requirements: The identified technical requirements are as follows:

1. A current of 450 A high pulse, 300–350 A average current.
2. No fluids allowed for cooling.
3. Radiation environment: 11,195 rad/h gamma radiation.
4. Remote tungsten change-out capability.
6. Tungsten with a diameter of 1/8 to 5/32 inch (3.1 to 3.9 mm).
7. Gases: 100% Ar; 95% Ar-5% H_2; or 75% Ar-25% He shielding gas composition. Exact shielding gas composition will be determined at a later date.
6. Adjustable gas cup.

Vendor 1: CK Worldwide Industries:

1. Does not have any torches
2. Is willing to develop a torch
3. Has sent non-solicited concept drawings
4. Does not have engineering analysis capabilities

Vendor 2: Weldcraft:

1. Does not have any torches
2. Torches with automatic tungsten changers have been used in Japan, but are not sold commercially

3. May address the design, but did not seem interested in assigning the development a very high priority
4. Has engineering analysis capabilities

Conclusion: There are no commercially-available auxiliary gas cooled welding torches that could be used by the Idaho National Engineering and Environmental Laboratory (INEEL) for this application. None of them appear to have remote change-out capabilities. Torches will have to be designed and developed by INEEL. The private industry sector may develop the design into a product to be purchased by INEEL.

Wire brush

Technical requirements: The identified technical requirements are as follows:

1. Radiation environment: 11,195 rad/h gamma radiation field
2. Operate in a narrow-groove weld profile
3. Operate over welds with a maximum interpass temperature of 350°F
4. Stainless steel wire

Vendor 1: Anderson Products:

1. Domestic supplier
2. Meets the above criteria

Vendor 2: Osborn International:

1. Domestic supplier
2. Meets the above criteria

Vendor 3: Weiler Corporation:

1. Domestic supplier
2. Meets the above criteria

Conclusion: The wire wheels required for this application are commercially available.

Grinding wheel

Technical requirements: The identified technical requirements are as follows:

1. Radiation environment: 11,195 rad/h gamma radiation field
2. Operate in a narrow-groove weld profile
3. Operate over welds with a maximum interpass temperature of 350°F
4. Aluminum oxide composition that is contaminate free (e.g., no iron, sulfur, or chlorine)

Vendor 1: Norton Abrasives:

1. Domestic supplier
2. Meets the above criteria

Conclusion: The grinding wheels needed for this application are commercially available.

Grinding head and motor

Technical requirements: The identified technical requirements are as follows:

1. Radiation environment: 11,195 rad/h gamma radiation field
2. Ability to use standard abrasive wheel arbors
3. Operate over welds with a maximum interpass temperature of 350°F

Vendor 1: RAD Robotic-Accessories:

1. Domestic supplier
2. Meets the above criteria

Vendor 2: PushCorp, Inc:

1. Domestic supplier
2. Meets the above criteria

Conclusion: Brush motors needed for this application are commercially available. However, the mounts may have to be modified to reach the surfaces.

Wire feeder

Technical requirements: The identified technical requirements are as follows:

1. Radiation environment: 11,195 rad/h gamma radiation field
2. Remote control capable
3. Ability to handle wire sizes of 0.035 in (0.88 mm) and 0.045 in (1.14 mm) diameter
4. Wire speed 25–300 inches per minute (0.63–7.59 m/min)
5. Electrical power required is 110 VAC

Vendor 1: AMET:

1. Domestic supplier
2. Meets the above criteria except the one for radiation

Vendor 2: Jetline Engineering, Inc.:

1. Domestic supplier
2. Meets the above criteria except the one for radiation

Vendor 3: Cyclomatic/Jetline Engineering, Inc.:

1. Domestic supplier
2. Meets the above criteria except the one for radiation

Conclusion: There are many commercial wire feeders available that would suit this application. However, all of the commercial units would need to be modified to operate in the expected radiation environment.

Wire guide positioners

Technical requirements: The identified technical requirements are as follows:

1. Radiation environment: 11,195 rad/h gamma radiation field
2. Two degrees of freedom
3. Remote control capability
4. Operate over welds with a maximum interpass temperature of 350°F
5. Wire guide adjustments to be 2-axis, horizontal and vertical, ±0.25 in (6.3 mm) each

Vendor 1: AMET:

1. Domestic supplier
2. Meets the above criteria

Vendor 2: Jetline Engineering, Inc.:

1. Domestic supplier
2. Meets the above criteria

Vendor 3: Panasonic:

1. Domestic supplier
2. Custom orders
3. Meets the above criteria

Conclusion: Commercial wire positioners are available. However, these units are fairly large in size and weight.

Wire cutting station

Technical requirements: The identified technical requirements are as follows:

1. Radiation environment: 11,195 rad/hr gamma radiation field
2. Ability to cut filler wire close to wire guide positioner
3. Ability to handle wire sizes of 0.035 in (0.88 mm) and 0.045 in (1.14 mm) diameter
4. Remote control capability

Vendor 1: Alexander Binzel Corporation:

1. Foreign supplier
2. Meets the above criteria

Vendor 2: Thermadyne:

1. Domestic supplier
2. Meets the above criteria

Vendor 3: Tregaskis:

1. Foreign supplier
2. Meets the above criteria

Conclusion: Commercial wire cutters were used on the GMAW robotic application. The stock commercial unit may have to be modified to meet the required configuration.

Arc viewing camera system

Technical requirements: The identified technical requirements are as follows:

1. Radiation environment: 11,195 rad/h gamma radiation field
2. Should last at least 60 hours of operation within specified radiation field
3. Provide clear view of welding pool and joint area
4. Operate over welds with a maximum interpass temperature of 350°F
5. 110 VAC input power

Vendor 1: AMI Arc Machines, Inc. "Direct Arc View Color Video System":

1. Domestic supplier.
2. AMI Arc Machines offers a high quality image system. This company has a good reputation in the welding industry. The system is

a charge-coupled device (CCD)-based camera system with auto switching between viewing modes. AMI Arc Machines does custom designs as part of daily business.

3. Does not meet the above criteria.

Vendor 2: Jetline Engineering, Inc. "JetView":

1. Domestic supplier.
2. Jetline Engineering offers a quality image system utilizing a simple arc suppression scheme. Jetline Engineering is a reputable company in the welding industry. The system is a CCD-based camera system with manual switching between viewing modes. Jetline Engineering does not generally perform custom designs far from its base system on arc viewing systems.
3. Does not meet the above criteria.

Vendor 3: Control Vision, Inc. "WeldingCam":

1. Domestic supplier.
2. Control Vision supplies a quality image system based on current technology. The system is a CCD-based camera system with the option to run it in an automatic switching mode. Control Vision does custom designs frequently. This company is very small, and it is only based on camera systems.
3. Does not meet the above criteria.

Vendor 4: Symphotic TII Corp "MegRAD 1":

1. Domestic supplier.
2. Symphotic TII Corp offers a radiation hardened camera system that can operate in the radiation fields. The system is not an arc-viewing camera system, but it may be possible to combine this camera with another manufacturer's arc-viewing technology to provide a radiation hardened arc-viewing camera system.
3. Does not meet the above criteria.

Conclusion: All of the arc-viewing camera systems surveyed provide images suitable for remote welding, and AMI Arc Machines is willing to modify symphotic camera to add arc suppression capabilities. One of the above arc-viewing camera manufacturing companies could build a camera system based on its existing design that would implement the MegRAD 1 camera, providing the radiation tolerance required. The custom design would also be required to provide cooling to meet the high temperature requirements.

Quantitative visual topography inspection

Technical requirements: The identified technical requirements are as follows:

1. Provide for seam tracking and weld profile measurements in a single package
2. Radiation environment: 11,195 rad/h gamma radiation field
3. Should last a minimum of 60 hours of operation within specified radiation field
4. Operate over welds with a maximum interpass temperature of 350°F
5. Distinguish a defect with a magnitude of 1/32 inch (0.8 mm)
6. About 2–3 Hz update rates for 10 ipm (25.3 cm/min) seam tracking
7. Operate with a cold wire feed tungsten inert gas welding system utilizing camera systems both in front and behind the weld torch
8. Input power of 110 VAC

Vendor 1: Servo Robot:

1. Foreign supplier. It has a domestic office, but its products are foreign-built in Canada.
2. This company builds a large line of products for laser based seam tracking and inspection. Servo Robot seems to be the leader in the field of this technology, and it has the widest array of products. The hardware utilizes complimentary metal oxide semiconductor (CMOS) based imagers which are much more radiation tolerant than the more common CCD imagers.

Vendor 2: Jetline Engineering, Inc.:

1. Domestic supplier.
2. This company builds a wide range of welding hardware. Jetline Engineering offers both a tactile and optical seam tracking system. Neither system appears to be well tailored to this application, and they do not provide any ability of doing weld profile measurements.

Vendor 3: Meta Vision Systems:

1. Foreign supplier (Canada).
2. This company builds a laser CCD camera seam tracking system. This vendor does not have any standard off-the-shelf systems for weld bead profile measuring, but it has done several custom systems. The vendor states that it could combine its seam tracking system with one of its previously-designed weld profile systems to provide a package meeting these requirements. Technically, this system would meet the DOE needs, but it is based on a charged-coupled device (CCD) camera that is inherently not radiation tolerant.

Vendor 4: Q-Tec Engineering:

1. Foreign supplier (Canada). This company builds a unique circular laser CCD camera based seam tracking system. The circular scan not only provides the ability to track the joint laterally, but it also provides the direction of the joint. While this functionality is unique to this sensor package, it is not needed for the DOE application. The increased complexity of the circular scanning laser that provides no increased value to this application makes this system unattractive for this project. The system is also based on a CCD camera that is inherently not radiation tolerant.

Conclusion: A system could be purchase-based on currently off-the-shelf hardware, which is customized to meet this specific application.

Eddy Current Testing (ET) equipment

Technical requirements: The identified technical requirements are as follows:

1. Provide inspection of weld surfaces
2. Operates over welds with a maximum surface temperature of 200°F
3. Radiation environment: 11,195 rad/h
4. Must be able to inspect fillet, repair cavity, and final weld surfaces
5. Remote operation
6. Evaluate software to established flaw criteria to ASME code requirements
7. 2–3 Hz update rates during inspection

Vendor 1: Agfa NDT, Inc.:

1. Foreign Vendor.
2. Supplies multi-frequency systems, but has limited support for multiple probes.
3. Serial interface for communication.
4. No array equipment.
5. Provides conventional and custom inspection probes. Development will be required.
6. No computer evaluation software.

Vendor 2: Jentek Sensors, Inc.:

1. Domestic Vendor
2. Supplies a multi-frequency system
3. Uses a meandering coil technique with a quantitative inversion computer algorithm
4. Coils are placed on kapton, which is moldable into the weld repair cavity
5. Probe to the control box is 15 to 30 ft (4.5 to 9 m), from the control box to the computer is 100 ft (30 m)
6. Probe development is required to address weld configurations
7. No computer evaluation software

Vendor 3: Quality Network, Inc.:

1. Foreign Vendor.
2. Grouped personal computer boards, up to 64 channels; all functions controlled via software.
3. Complete executable software or C++ available, no Lab View© (for board operation).
4. Has built-in probes (little information available). Can use almost any available probe, but will require development.
5. Sensor-to-equipment distance: 15 ft (4.5 m).
6. Not advertised as an array system. However, with software development it could provide array capability.
7. No computer evaluation software.

Vendor 4: RD-Tech:

1. Foreign Vendor. It is Canadian owned, but it has U.S. representatives and facilities.
2. MS 5800, multi-frequency system, supports array or multiple channel probe scanning.
3. Stand-alone system, various communication ports to system computer.
4. Test frequency 20 Hz to 6 MHz.
5. System communicates to acquisition computer system using a 100 Base-T Fast Ethernet.
6. Some thresholding analysis software available. It offers software a development kit.
7. Offers special moldable probes for repair cavities; also offers array probes that are 10 to 40 times faster than a rastered single probe.
8. Development and integration work will have to be done.

Vendor 5: Staveley NDT Technologies:

1. Domestic Vendor
2. Power station can support up to 4 probes, 100 Hz to 12 MHz
3. System controlled via serial line; eddy current output is analog

4. Instrument-to-probe distance is 25 ft (7.5 m)
5. Development and integration work will have to be done
6. There are some questions on probe ability to withstand radiation fluxes

Vendor 6: ZETEC, Inc.:

1. Domestic Vendor
2. Multi-frequency system that can support up to 40 probes
3. System communicates to acquisition computer system using a 100 Base-T Fast Ethernet
4. Offers conventional as well as custom probes

Conclusion: The commercially-available equipment meets some of the application needs for the eddy current system. However, integration and probe development will be required.

Ultrasonic nondestructive examination (NDE) equipment

Technical requirements: The identified technical requirements are as follows:

1. Remote post and in-process inspection required
2. Surface temperature: 200°F
3. Either uses no couplant or limited amounts of deionized water
4. Radiation environment: 11,195 Rad/h
5. Inspection speeds = 8 ipm (20.2 cm/min), minimum

Table 5 contains a summary of the vendor capabilities. Network Interface Card (NIC) testing is a mature technology with commercial vendors supplying a broad range of transducers, instrumentation, and integrated systems. However, for this application, the stated requirements stretch or exceed the technical capabilities of the existing technologies:

Table 5. Vendors surveyed (Watkins 2003a).

Vendors	Conventional Transducers	Arrays	Wheel Probes	EMAT	Instrumentation	Systems	Domestic	Comments
Dapco Industries Inc.	X		X			X	X	Supplies transducers and specialized systems; wheel probes requires couplant to function
AGFA - KrautKramer	X	X			X	X		Supplies phased array transducers and systems
GE - Panametrics	X			X	X	X	X	Supplies conventional transducers and instrumentation
RD - Tech	X	X		X	X	X		Supplies phased array transducers and systems; water wedge technology
Sigma Transducers	X	X	X				X	Supplies conventional transducers, dry coupled wheel probes, and flexible arrays
Sonatest Inc.	X		X		X	X		Supplies conventional transducers and dry coupled wheel probes; wheel probes cannot tolerate 200°F
Staveley NDT Technologies	X		X		X	X	X	Supplies conventional transducers, dry coupled wheel probes, and field systems; wheel probes cannot tolerate 200°F
NDT International	X		X		X	X	X	Supplies conventional transducers, dry coupled wheel probes, and handheld field systems
Sonix					X	X	X	Supplies scanning systems and software
UTEX Scientific Instruments Inc.					X	X		Supplies system components and scanning software

Table 5. (cont'd).

Vendors	Conventional Transducers	Arrays	Wheel Probes	EMAT	Instrumentation	Systems	Domestic	Comments
Harfang Microtechniques Inc.							X	Supplies handheld phased array system
Imasonic	X	X						Supplies conventional and phased array transducers
EMAT Ultrasonics Inc.				X		X	X	Supplies EMATS and systems
Innerspec Technologies Inc.				X		X	X	Supplies EMATS and systems
Matec Instruments Companies Inc.					X	X	X	Supplies high power systems for conventional ultrasonics and EMATS
Ritec Inc.					X	X	X	Supplies high power systems for conventional ultrasonics and EMATS

1. Surface Temperature 200°F (transients up to 930°F): High-temperature transducers exist that can tolerate the 200-930°F environment. These transducers typically require high temperature couplants and intermittent contact. Non-contacting transducers, such as Electromagnetic Acoustic Transducers (EMATS), are better suited for continuous high-temperature testing. They require no couplant and can be built to work continuously at 700°F with transients to 1,300°F. Unfortunately, EMATS suffer from low-transmitted ultrasonic energy, thereby resulting in low signal to noise ratios. Cooling can be provided to the transducer to keep the transducer from degrading due to temperature.

2. No appreciable liquid couplant: Conventional ultrasonic techniques use water or a similar liquid/gel material to mechanically couple the ultrasonic energy into the test sample. Obviously, the use of liquid couplant is not always viable, and techniques have been developed to limit or remove the need for couplant. Those techniques include dry coupling, electromagnetics, and lasers. Dry coupling depends on soft elastomers and pressure to couple the transducer to the test piece. The limitations of this approach are that the elastomers significantly attenuate the ultrasonic energy (yielding low signal to noise ratios), and are susceptible to both heat and radiation. Electromagnetic techniques use magnetic fields to excite and detect the sound. This approach is non-contacting, and it is well suited for elevated-temperature testing; however, due to limited coupling, it suffers from low-ultrasonic energy transmission. Laser techniques use high-energy laser pulses to produce ultrasound and interferometric techniques to detect the surface vibrations of the return signals. Laser-based technologies are less developed than the other techniques, and they suffer from low detection efficiency; sensitivity to surface conditions; and complexity. It is also more difficult to control beam focusing and directionality. Membranes, which seep small amounts of water to create a water film for coupling to the inspection surface, are not always commercially available.

3. Radiation environment: Ultrasonic techniques have been successfully used in radiation environments, but test times have been short. As a result, the long-term durability of the transducers and associated cabling in high-radiation fields is not known.

4. Welding speeds = 8 ipm (20.2 cm/min): Inspection rates are more than adequate to address an 8 ipm (20.2 cm/min) welding speed.

Conclusion: Instrumentation and systems exist to meet the basic requirements. Although systems and scanning software exist, work will need to be performed to integrate/interface the systems and software with the robotics and process control system. Probe assemblies are not commercially available to fully address the lack of coupling, temperature, and radiation fields. Work will need to be performed to develop and test appropriate sensors. It will also be necessary to design transducer configurations to match the specific test geometry for this application.

Linear slides/positioning actuators

These are components of an automated/remote welding and inspection system for positioning of various components of the system (e.g., weld torch, probes), having the following technical requirements:

1. Provide positioning of weld torch, repair groove grinder, and nondestructive evaluation probes
2. Operate over welds with a maximum surface temperature of 350°F
3. Radiation environment: 11,195 rad/h
4. Remote operation

Vendor: Parker Hannifin:

1. Domestic supplier
2. Slides available with radiation-hardened parts to meet technical requirements

Conclusion: The commercially-available slides meet DOE requirements.

Commercial center pivot weld systems

Technical requirements: The identified technical requirements are as follows:

1. Radiation environment: 11,195 rad/h
2. Remote operation
3. Nondestructive examination (NDE) inspection
4. Repair capabilities

Vendor 1: Berkeley Process Controls:

1. Domestic supplier
2. Center pivot is placed on the lid and uses the canister as a base
3. Designed for topside lid weld; would need to be customized for the canister
4. Welding only; no remote NDE or welding capabilities
5. No repair capabilities

Vendor 2: AMI Arc Machines, Inc.:

1. Center pivot is placed on the lid and uses the canister as a base
2. Designed for topside lid weld; would need to be customized for the canister
3. Welding only; no remote NDE or welding capabilities
4. No repair capabilities

Vendor 3: AMET:

1. Center pivot is placed on the lid and uses the canister as a base
2. Designed for topside lid weld; would need to be customized for the canister
3. Welding only; no remote NDE or welding capabilities
4. No repair capabilities

Conclusion: Complete systems with the necessary capabilities are not currently commercially available. Components could be bought that are commercially available for parts of the system. These components would have to be integrated into the designed weld and inspection system.

Legal and Related Requirements

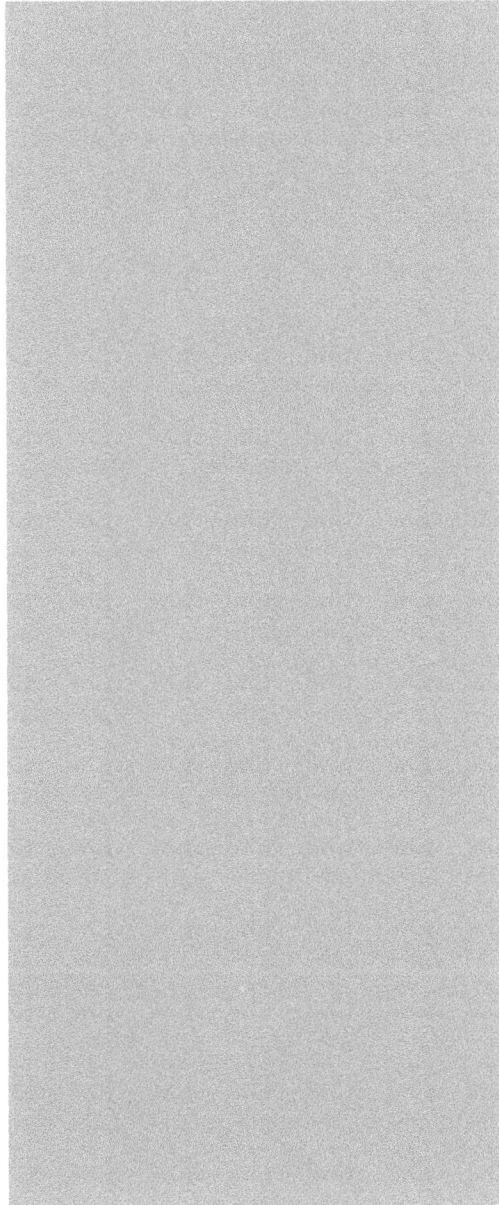

The development of standardized spent nuclear fuel canisters for the U.S. Department of Energy requires compliance with a number of laws, regulations and standards. This chapter contains information on these requirements.

THE NUCLEAR WASTE POLICY ACT AS AMENDED

This section contains excerpts from the Nuclear Waste Policy act (Public Law 97-425; 96 Stat. 2201), as amended by PL. 100-203, Title V, Subtitle A (December 22, 1987), PL. 100-507 (October 18, 1988), and PL. 102-486 (The Energy Policy Act of 1992, October 24, 1992). The Act is codified at 42 U.S.C. 10101 et seq.

THE NUCLEAR WASTE POLICY ACT OF 1982

An Act to provide for the development of repositories for the disposal of high-level radioactive waste and spent nuclear fuel, to establish a program of research, development, and demonstration regarding the disposal of high-level radioactive waste and spent nuclear fuel, and for other purposes. [. . .]

DEFINITIONS

Sec. 2. For purposes of this Act

(1) The term "Administrator" means the Administrator of the Environmental Protection Agency. [. . .]
(3) The term "atomic energy defense activity" means any activity of the Secretary [of Department of Energy] performed in whole or in part in carrying out any of the following functions:

    (A) naval reactors development;

    (B) weapons activities including defense inertial confinement fusion;

    (C) verification and control technology;

    (D) defense nuclear materials production;

    (E) defense nuclear waste and materials by-products management;

    (F) defense nuclear materials security and safeguards and security investigations; and

    (G) defense research and development. [. . .]

(5) The term "civilian nuclear activity" means any atomic energy activity other than an atomic energy defense activity.

(6) The term "civilian nuclear power reactor" means a civilian nuclear powerplant required to be licensed under section 103 or 104 b. of the Atomic Energy Act of 1954 [42 U.S.C. 2133, 2134(b)].

(7) The term "Commission" means the Nuclear Regulatory Commission.

(8) The term "Department" means the Department of Energy.

(9) The term "disposal" means the emplacement in a repository of high-level radioactive waste, spent nuclear fuel, or other highly radioactive material with no foreseeable intent of recovery, whether or not such emplacement permits the recovery of such waste.

(10) The terms "disposal package" and "package" mean the primary container that holds, and is in contact with, solidified high-level radioactive waste, spent nuclear fuel, or other radioactive materials, and any overpacks that are emplaced at a repository.

(11) The term "engineered barriers" means manmade components of a disposal system designed to prevent the release of radionuclides into the geologic medium involved. Such term includes the high-level radioactive waste form, high-level radioactive waste canisters, and other materials placed over and around such canisters.

(12) The term "high-level radioactive waste" means—

  (A) the highly radioactive material resulting from the reprocessing of spent nuclear fuel, including liquid waste produced directly in reprocessing and any solid material derived from such liquid waste that contains fission products in sufficient concentrations; and

  (B) other highly radioactive material that the Commission, consistent with existing law, determines by rule requires permanent isolation. [. . .]

(23) The term "spent nuclear fuel" means fuel that has been withdrawn from a nuclear reactor following irradiation, the constituent elements of which have not been separated by reprocessing. [. . .]

(25) The term "storage" means retention of high-level radioactive waste, spent nuclear fuel, or transuranic waste with the intent to recover such waste or fuel for subsequent use, processing, or disposal. [. . .]

TRANSPORTATION

Sec. 137. (a)

(1) Transportation of spent nuclear fuel under section 136(a) [42 U.S.C. 10136(a)] shall be subject to licensing and regulation by the Commission

and by the Secretary of Transportation as provided for transportation of commercial spent nuclear fuel under existing law.

(2) The Secretary, in providing for the transportation of spent nuclear fuel under this Act [42 U.S.C. 10101 et seq.], shall utilize by contract private industry to the fullest extent possible in each aspect of such transportation. The Secretary shall use direct Federal services for such transportation only upon a determination of the Secretary of Transportation, in consultation with the Secretary, that private industry is unable or unwilling to provide such transportation services at reasonable cost. [42 U.S.C. 10157]

SUBTITLE C—MONITORED RETRIEVABLE STORAGE
MONITORED RETRIEVABLE STORAGE

Sec. 141. (a) Findings. The Congress finds that—

(1) long-term storage of high-level radioactive waste or spent nuclear fuel in monitored retrievable storage facilities is an option for providing safe and reliable management of such waste or spent fuel;
(2) the executive branch and the Congress should proceed as expeditiously as possible to consider fully a proposal for construction of one or more monitored retrievable storage facilities to provide such long-term storage;
(3) the Federal Government has the responsibility to ensure that site-specific designs for such facilities are available as provided in this section;
(4) the generators and owners of the high-level radioactive waste and spent nuclear fuel to be stored in such facilities have the responsibility to pay the costs of the long-term storage of such waste and spent fuel; and
(5) disposal of high-level radioactive waste and spent nuclear fuel in a repository developed under this Act [42 U.S.C. 10101 et seq.] should proceed regardless of any construction of a monitored retrievable storage facility pursuant to this section. [. . .]

REGULATION OF U.S. NUCLEAR REGULATORY COMMISSION ON DISPOSAL OF HIGH-LEVEL RADIOACTIVE WASTES IN GEOLOGIC REPOSITORIES

This section contains excerpts from regulations dealing with disposal of high-level radioactive waste (HLW) promulgated by the U.S. Nuclear Regulatory Commission and codified as 10 CFR PART 60.

§60.43 License specification.

(a) A license issued under this part shall include license conditions derived from the analyses and evaluations included in the application, including amendments made before a license is issued, together with such additional conditions as the Commission finds appropriate.
(b) License conditions shall include items in the following categories:

(1) Restrictions as to the physical and chemical form and radio-isotopic content of radioactive waste.

(2) Restrictions as to size, shape, and materials and methods of construction of radioactive waste packaging.

(3) Restrictions as to the amount of waste permitted per unit volume of storage space considering the physical characteristics of both the waste and the host rock.

(4) Requirements relating to test, calibration, or inspection to assure that the foregoing restrictions are observed.

(5) Controls to be applied to restricted access and to avoid disturbance to the postclosure controlled area and to areas outside the controlled area where conditions may affect isolation within the controlled area.

(6) Administrative controls, which are the provisions relating to organization and management, procedures, recordkeeping, review and audit, and reporting necessary to assure that activities at the facility are conducted in a safe manner and in conformity with the other license specifications. [. . .]

§60.113 Performance of particular barriers after permanent closure.

(a) General provisions—(1) Engineered barrier system.

(i) The engineered barrier system shall be designed so that assuming anticipated processes and events:

(A) Containment of HLW will be substantially complete during the period when radiation and thermal conditions in the engineered barrier system are dominated by fission product decay; and

(B) any release of radionuclides from the engineered barrier system shall be a gradual process which results in small fractional releases to the geologic setting over long times. For disposal in the saturated zone, both the partial and complete filling with groundwater of available void spaces in the underground facility shall be appropriately considered and analysed among the anticipated processes and events in designing the engineered barrier system.

(ii) In satisfying the preceding requirement, the engineered barrier system shall be designed, assuming anticipated processes and events, so that:

(A) Containment of HLW within the waste packages will be substantially complete for a period to be determined by the Commission talcing into account the factors specified in §60.113(b) provided, that such period shall be not less than 300 years nor more than 1,000 years after permanent closure of the geologic repository; and

(B) The release rate of any radionuclide from the engineered barrier system following the containment period shall not exceed one part in 100,000 per year of the inventory of that radionuclide calculated to be present at 1,000 years following permanent closure, or such other fraction of the inventory as may be approved or specified by the Commission; provided, that this requirement does not apply to any radionuclide which is released at a rate less than 0.1% of the calculated total release rate limit. The calculated total release rate limit shall be taken to be one part in 100,000 per year of the inventory of radioactive waste, originally emplaced in the underground facility, that remains after 1,000 years of radioactive decay.

§60.131 General design criteria for the geologic repository operations area.

(a) Radiological protection. The geologic repository operations area shall be designed to maintain radiation doses, levels, and concentrations of radioactive material in air in restricted areas within the limits specified in part 20 of this chapter. Design shall include:

(1) Means to limit concentrations of radioactive material in air;

(2) Means to limit the time required to perform work in the vicinity of radioactive materials, including, as appropriate, designing equipment

for ease of repair and replacement and providing adequate space for ease of operation;

(3) Suitable shielding;

(4) Means to monitor and control the dispersal of radioactive contamination;

(5) Means to control access to high radiation areas or airborne radioactivity areas; and

(6) A radiation alarm system to warn of significant increases in radiation levels, concentrations of radioactive material in air, and of increased radioactivity released in effluents. The alarm system shall be designed with provisions for calibration and for testing its operability.

(b) Protection against design basis events. The structures, systems, and components important to safety shall be designed so that they will perform their necessary safety functions, assuming occurrence of design basis events.

(c) Protection against dynamic effects of equipment failure and similar events. The structures, systems, and components important to safety shall be designed to withstand dynamic effects such as missile impacts, that could result from equipment failure, and similar events and conditions that could lead to loss of their safety functions.

(d) Protection against fires and explosions.

(1) The structures, systems, and components important to safety shall be designed to perform their safety functions during and after credible fires or explosions in the geologic repository operations area.

(2) To the extent practicable, the geologic repository operations area shall be designed to incorporate the use of noncombustible and heat resistant materials.

(3) The geologic repository operations area shall be designed to include explosion and fire detection alarm systems and appropriate suppression systems with sufficient capacity and capability to reduce the adverse effects of fires and explosions on structures, systems, and components important to safety.

(4) The geologic repository operations area shall be designed to include means to protect systems, structures, and components important to safety against the adverse effects of either the operation or failure of the fire suppression systems.

(e) Emergency capability.

(1) The structures, systems, and components important to safety shall be designed to maintain control of radioactive waste and radioactive

effluents, and permit prompt termination of operations and evacuation of personnel during an emergency.

(2) The geologic repository operations area shall be designed to include onsite facilities and services that ensure a safe and timely response to emergency conditions and that facilitate the use of available offsite services (such as fire, police, medical, and ambulance service) that may aid in recovery from emergencies.

(f) Utility services.

(1) Each utility service system that is important to safety shall be designed so that essential safety functions can be performed, assuming occurrence of the design basis events.

(2) The utility services important to safety shall include redundant systems to the extent necessary to maintain, with adequate capacity, the ability to perform their safety functions.

(3) Provisions shall be made so that, if there is a loss of the primary electric power source or circuit, reliable and timely emergency power can be provided to instruments, utility service systems, and operating systems, including alarm systems, important to safety.

(g) Inspection, testing, and maintenance. The structures, systems, and components important to safety shall be designed to permit periodic inspection, testing, and maintenance, as necessary, to ensure their continued functioning and readiness.

(h) Criticality control. All systems for processing, transporting, handling, storage, retrieval, emplacement, and isolation of radioactive waste shall be designed to ensure that nuclear criticality is not possible unless at least two unlikely, independent, and concurrent or sequential changes have occurred in the conditions essential to nuclear criticality safety. Each system must be designed for criticality safety assuming occurrence of design basis events. The calculated effective multiplication factor (k_{eff}) must be sufficiently below unity to show at least a 5 percent margin, after allowance for the bias in the method of calculation and the uncertainty in the experiments used to validate the method of calculation.

(i) Instrumentation and control systems. The design shall include provisions for instrumentation and control systems to monitor and control the behavior of systems important to safety, assuming occurrence of design basis events.

(j) Compliance with mining regulations. To the extent that DOE is not subject to the Federal Mine Safety and Health Act of 1977, as to the construction and operation of the geologic repository operations area, the design of the geologic repository operations area shall nevertheless include provisions for worker protection necessary to provide reasonable assurance that all structures, systems, and components important to safety can perform their intended functions. Any deviation from relevant design requirements in 30 CFR, chapter I, subchapters D, E, and N will give rise to a rebuttable presumption that this requirement has not been met.

(k) Shaft conveyances used in radioactive waste handling.

(1) Hoists important to safety shall be designed to preclude cage free fall.

(2) Hoists important to safety shall be designed with a reliable cage location system.

(3) Loading and unloading systems for hoists important to safety shall be designed with a reliable system of interlocks that will fail safely upon malfunction.

(4) Hoists important to safety shall be designed to include two independent indicators to indicate when waste packages are in place and ready for transfer. [. . .]

§60.130 General considerations.

Pursuant to the provisions of §60.21(c)(2)(i), an application to receive, possess, store, and dispose of high-level radioactive waste in the geologic repository operations area must include the principal design criteria for a proposed facility. The principal design criteria establish the necessary design, fabrication, construction, testing, maintenance, and performance requirements for structures, systems, and components important to safety and/or important to waste isolation. Sections 60.131 through 60.134 specify minimum requirements for the principal design criteria for the geologic repository operations area.

These design criteria are not intended to be exhaustive. However, omissions in §§60.131 through 60.134 do not relieve DOE from any obligation to provide such features in a specific facility needed to achieve the performance objectives. [. . .]

Design Criteria for the Geological Repository Operations Area

§60.133 Additional design criteria for the underground facility.

(a) General criteria for the underground facility.

(1) The orientation, geometry, layout, and depth of the underground facility, and the design of any engineered barriers that are part of the underground facility shall contribute to the containment and isolation of radionuclides.

(2) The underground facility shall be designed so that the effects of credible disruptive events during the period of operations, such as flooding, fires and explosions, will not spread through the facility.

(b) Flexibility of design. The underground facility shall be designed with sufficient flexibility to allow adjustments where necessary to accommodate specific site conditions identified through in situ monitoring, testing, or excavation.

(c) Retrieval of waste. The underground facility shall be designed to permit retrieval of waste in accordance with the performance objectives of §§60.111.

(d) Control of water and gas. The design of the underground facility shall provide for control of water or gas intrusion.

(e) Underground openings.

(1) Openings in the underground facility shall be designed so that operations can be carried out safely and the retrievability option maintained.

(2) Openings in the underground facility shall be designed to reduce the potential for deleterious rock movement or fracturing of overlying or surrounding rock.

(f) Rock excavation. The design of the underground facility shall incorporate excavation methods that will limit the potential for creating a preferential pathway for groundwater to contact the waste packages or radionuclide migration to the accessible environment.

(g) Underground facility ventilation. The ventilation system shall be designed to:

(1) Control the transport of radioactive particulates and gases within and releases from the underground facility in accordance with the performance objectives of §§60.111(a),

(2) Assure the ability to perform essential safety functions assuming occurrence of design basis events.

(3) Separate the ventilation of excavation and waste emplacement areas.

(h) Engineered barriers. Engineered barriers shall be designed to assist the geologic setting in meeting the performance objectives for the period following permanent closure.

(i) Thermal loads. The underground facility shall be designed so that the performance objectives will be met taking into account the predicted thermal and thermo mechanical response of the host rock, and surrounding strata, groundwater system. [. . .]

§60.135 Criteria for the waste package and its components.

(a) High-level-waste package design in general.

(1) Packages for HLW shall be designed so that the in situ chemical, physical, and nuclear properties of the waste package and its interactions with the emplacement environment do not compromise the function of the waste packages or the performance of the underground facility or the geologic setting.

(2) The design shall include but not be limited to consideration of the following factors: solubility, oxidation/reduction reactions, corrosion, hydriding, gas generation, thermal effects, mechanical strength, mechanical stress, radiolysis, radiation damage, radionuclide retardation, leaching, fire and explosion hazards, thermal loads, and synergistic interactions.

(b) Specific criteria for HLW package design—

(1) Explosive, pyrophoric, and chemically reactive materials. The waste package shall not contain explosive or pyrophoric materials or chemically reactive materials in an amount that could compromise the ability of the underground facility to contribute to waste isolation or the ability of the geologic repository to satisfy the performance objectives.

(2) Free liquids. The waste package shall not contain free liquids in an amount that could compromise the ability of the waste packages to achieve the performance objectives relating to containment of HLW (because of chemical interactions or formation of pressurized vapor) or result in spillage and spread of contamination in the event of waste package perforation during the period through permanent closure.

(3) Handling. Waste packages shall be designed to maintain waste containment during transportation, emplacement, and retrieval.

(4) Unique identification. A label or other means of identification shall be provided for each waste package. The identification shall not impair the integrity of the waste package and shall be applied in such a way that the information shall be legible at least to the end of the period of retrievability. Each waste package identification shall be consistent with the waste package's permanent written records.

(c) Waste form criteria for HLW. High-level radioactive waste that is emplaced in the underground facility shall be designed to meet the following criteria:

(1) Solidification. All such radioactive wastes shall be in solid form and placed in sealed containers.

(2) Consolidation. Particulate waste forms shall be consolidated (for example, by incorporation into an encapsulating matrix) to limit the availability and generation of particulates.

(3) Combustibles. All combustible radioactive wastes shall be reduced to a noncombustible form unless it can be demonstrated that a fire involving the waste packages containing combustibles will not compromise the integrity of other waste packages, adversely affect any structures, systems, or components important to safety, or compromise the ability of the underground facility to contribute to waste isolation.

(d) Design criteria for other radioactive wastes. Design criteria for waste types other than HLW will be addressed on an individual basis if and when they are proposed for disposal in a geologic repository.

REGULATION OF U.S. NUCLEAR REGULATORY COMMISSION ON PACKAGING AND TRANSPORTATION OF RADIOACTIVE MATERIAL

This section contains excerpts from regulations on packaging and transportation of radioactive materials promulgated by the U.S. Nuclear Regulatory Commission and codified as 10 CFR PART 71.

§71.4 Definitions.

Package means the packaging together with its radioactive contents as presented for transport.

Subpart D—Application for Package Approval

§71.35 Package evaluation.

The application must include the following:

(a) A demonstration that the package satisfies the standards specified in subparts E and F of this part;
(b) For a fissile material package, the allowable number of packages that may be transported in the same vehicle in accordance with §71.59; and
(c) For a fissile material shipment, any proposed special controls and precautions for transport, loading, unloading, and handling and any proposed special controls in case of an accident or delay. [. . .]

§71.43 General standards for all packages.

(a) The smallest overall dimension of a package may not be less than 10 cm (4 in).
(b) The outside of a package must incorporate a feature, such as a seal, that is not readily breakable and that, while intact, would be evidence that the package has not been opened by unauthorized persons.
(c) Each package must include a containment system securely closed by a positive fastening device that cannot be opened unintentionally or by a pressure that may arise within the package.

(d) A package must be made of materials and construction that assure that there will be no significant chemical, galvanic, or other reaction among the packaging components, among package contents, or between the packaging components and the package contents, including possible reaction resulting from in leakage of water, to the maximum credible extent. Account must be taken of the behavior of materials under irradiation.

(e) A package valve or other device, the failure of which would allow radioactive contents to escape, must be protected against unauthorized operation and, except for a pressure relief device, must be provided with an enclosure to retain any leakage.

(f) A package must be designed, constructed, and prepared for shipment so that under the tests specified in §71.71 ("Normal conditions of transport") there would be no loss or dispersal of radioactive contents, no significant increase in external surface radiation levels, and no substantial reduction in the effectiveness of the packaging.

(g) A package must be designed, constructed, and prepared for transport so that in still air at 38°C (100°F) and in the shade, no accessible surface of a package would have a temperature exceeding 50°C (122°F) in a non-exclusive use shipment, or 85°C (185°F) in an exclusive use shipment.

(h) A package may not incorporate a feature intended to allow continuous venting during transport. [. . .]

§71.55 General requirements for fissile material packages.

(a) A package used for the shipment of fissile material must be designed and constructed in accordance with §§71.41 through 71.47. When required by the total amount of radioactive material, a package used for the shipment of fissile material must also be designed and constructed in accordance with §71.51.

(b) Except as provided in paragraph (c) of this section, a package used for the shipment of fissile material must be so designed and constructed and its contents so limited that it would be subcritical if water were to leak into the containment system, or liquid contents were to leak out of the containment system so that, under the following conditions, maximum reactivity of the fissile material would be attained:

(1) The most reactive credible configuration consistent with the chemical and physical form of the material;

(2) Moderation by water to the most reactive credible extent; and

(3) Close full reflection of the containment system by water on all sides, or such greater reflection of the containment system as may additionally be provided by the surrounding material of the packaging.

(c) The Commission may approve exceptions to the requirements of paragraph (b) of this section if the package incorporates special design features that ensure that no single packaging error would permit leakage, and if appropriate measures are taken before each shipment to ensure that the containment system does not leak.

(d) A package used for the shipment of fissile material must be so designed and constructed and its contents so limited that under the tests specified in §71.71 ("Normal conditions of transport")—

(1) The contents would be subcritical;

(2) The geometric form of the package contents would not be substantially altered;

(3) There would be no leakage of water into the containment system unless, in the evaluation of undamaged packages under §71.59(a)(1), it has been assumed that moderation is present to such an extent as to cause maximum reactivity consistent with the chemical and physical form of the material; and

(4) There will be no substantial reduction in the effectiveness of the packaging, including:

(i) No more than 5 percent reduction in the total effective volume of the packaging on which nuclear safety is assessed;

(ii) No more than 5 percent reduction in the effective spacing between the fissile contents and the outer surface of the packaging; and

(iii) No occurrence of an aperture in the outer surface of the packaging large enough to permit the entry of a 10 cm (4 in) cube.

(e) A package used for the shipment of fissile material must be so designed and constructed and its contents so limited that under the tests specified in §71.73 ("Hypothetical accident conditions"), the package would be subcritical. For this determination, it must be assumed that:

(1) The fissile material is in the most reactive credible configuration consistent with the damaged condition of the package and the chemical and physical form of the contents;

(2) Water moderation occurs to the most reactive credible extent consistent with the damaged condition of the package and the chemical and physical form of the contents; and

(3) There is full reflection by water on all sides, as close as is consistent with the damaged condition of the package.

REGULATION OF U.S. NUCLEAR REGULATORY COMMISSION ON LICENSING REQUIREMENTS FOR THE INDEPENDENT STORAGE OF CERTAIN WASTES

This section contains an excerpt from regulations on licensing requirements for the independent storage of spent nuclear fuel, high-level radioactive waste, and reactor-related radioactive class C waste on promulgated by the U.S. Nuclear Regulatory Commission and codified as 10 CFR PART 72.

72.3 Definitions.

High-level radioactive waste or HLW means (1) the highly radioactive material resulting from the reprocessing of spent nuclear fuel, including liquid waste produced directly in reprocessing and any solid material derived from such liquid waste that contains fission products in sufficient concentrations; and (2) other highly radioactive material that the Commission, consistent with existing law, determines by rule requires permanent isolation.

CODES AND STANDARDS OF THE
AMERICAN SOCIETY OF MECHANICAL ENGINEERS

The American Society of Mechanical Engineers (ASME), also known as ASME International is a major source for a number of industrial codes of standards. Numerous ASME coded and standards have been adopted by the U.S. government agencies by reference and are an integral part of enforceable regulations.

Codes and standards dealing with spent nuclear fuel canisters and related issues emerged from Pressure Vessel Codes, initially developed for fossil fuel power plants. Although many sections of ASME codes and standards are applicable to all relevant installation, nuclear-specific issues are included Section III of the Boiler and Pressure Vessel (B&PV) Code. In addition, the ASME Nuclear Accreditation Program assures strict conformance with the ASME Codes and Standards for manufacturing of a nuclear facility. It covers the quality assurance of construction materials, design, engineering, operation, inspection, and continuing maintenance of the site. The purpose of the Accreditation program is to ascertain that the applicant has, uses and abides by a quality assurance manual which is clear and understandable.

Nuclear Stamp

As in other areas of activities of the ASME Codes and Standards, Code Symbol Stamps are used to indicate that the stamped items conform to the ASME Code. The Stamp provides confidence that the stamped items conform to established safety standards.

An organization receives a Certificate of Accreditation when it is recognized that it has an acceptable quality assurance program. However, no stamp can be issued until ASME has conducted an implementation audit for conversion to a Certificate of Authorization which authorizes the use of one or more of the five Code Symbol Stamps.

N - Nuclear vessels, pumps, valves, piping systems, storage tanks, core support structures, concrete containments, and transport packaging.

NA - Field installation and shop assembly.
NPT - Fabrication, with or without design responsibility, for nuclear appurtenances and supports.
NV - Pressure relief valves.
N3 - Containment for spent fuel and high level radioactive waste.

Content of Certain B&PV Code Sections

Section II. Materials:

This Section consists of four parts. Each part is a service book to the other Code Sections.

Part A - provides material specifications for ferrous materials adequate for safety in the field of pressure equipment. These specifications contain requirements and mechanical properties, test specimens, and methods of testing. Covered are steel pipe; steel tubes: fittings, valves; steel plates; wrought iron as examples.

Part B - covers nonferrous materials adequate for safety in the field of pressure equipment. These specifications contain requirements for heat treatment, manufacture, chemical composition, heat and product analyses, test specimens and methods of testing. Covered as examples are; aluminum, copper, nickel, titanium, and zirconium, all in alloys and in bars plates, and rods.

Part C - covers specifications for welding rods, electrodes, and filler materials. Provided are specifications for manufacture, chemical composition, mechanical usability, testing requirements and procedures, and intended use. As examples the requirements are for steel and steel alloy rod and electrodes; aluminum and aluminum base alloy rods, copper, nickel titanium and zirconium and their alloys in rods and electrodes.

Part D - contains tables of design stress values, tensile and yield strength values, and tables and charts of material values. Part D facilitates ready identification of specific materials to specific Sections of the B&PV code.

Subsection SA-530 - Specifications for General Requirements for Specialized Carbon and Alloy Steel Pipe.

This specification covers a group of requirements which are mandatory requirements to the ASTM pipe product specifications. The specifications cover manufacturing process; chemical composition; mechanical requirements (such as method of mechanical testing); tensile requirements; permissible variations in weight, wall thickness, inside and outside diameters, length, and repairs by welding; test specimens, hydrostatic tests requirements; inspection, rejection and marking.

Section III. Rules for Construction of Nuclear Facility Components

The basic coverage of this Section is the providing of requirements for the materials, design, fabrication, testing, inspection, installation, certification, stamping, and overpressure protection of nuclear power plant components, piping supports, metal vessels and systems, pumps and valves. There are three divisions in Section III.

Division 1: This division has seven subsections. Each subsection contains the requirements for the material, design, fabrication, inspection, testing and overpressure protection of items in different categories.
Subsection NB - Class 1 components for Class 1 construction.
Subsection NC - Class 2 components for Class 2 construction.
Subsection ND - Class 3 components for Class 3 construction.
Subsection NE - Class MC components for metal containment vessels for Class MNC construction.
Subsection NF - Supports for Classes 1,2,3, and MC construction.
Subsection NG - Core Support structures are those structures which provide direct support or restraint of the core within the reactor pressure vessel.
Subsection NH - Class 1 components in elevated temperature service.
Appendices - Appendices, both mandatory and nonmandatory for Section III, Division 1, including a listing of design and design analysis methods and information and data report forms.

Division 2: This division is the Code for concrete reactor vessels and containment.

Division 3: This division constitutes the requirements for the design and construction of the containment system of a nuclear spent fuel or high level radioactive waste transport packaging.

Section V - Nondestructive Examination

This section contains the requirements and methods for nondestructive examination which are referenced and required by other Code Sections. Included are manufacturers examination responsibilities, duties of authorized inspectors and requirements for qualification of personnel, inspection and examination. The requirements cover examination by radiographic, ultrasonic, liquid penetrant, and magnetic particle eddy current methods; visual examination; leak testing; and acoustic emission examination of fiber-reinforced plastic pressure vessels.

Peer Review Criteria, Findings, and Recommendations of the Review Panel

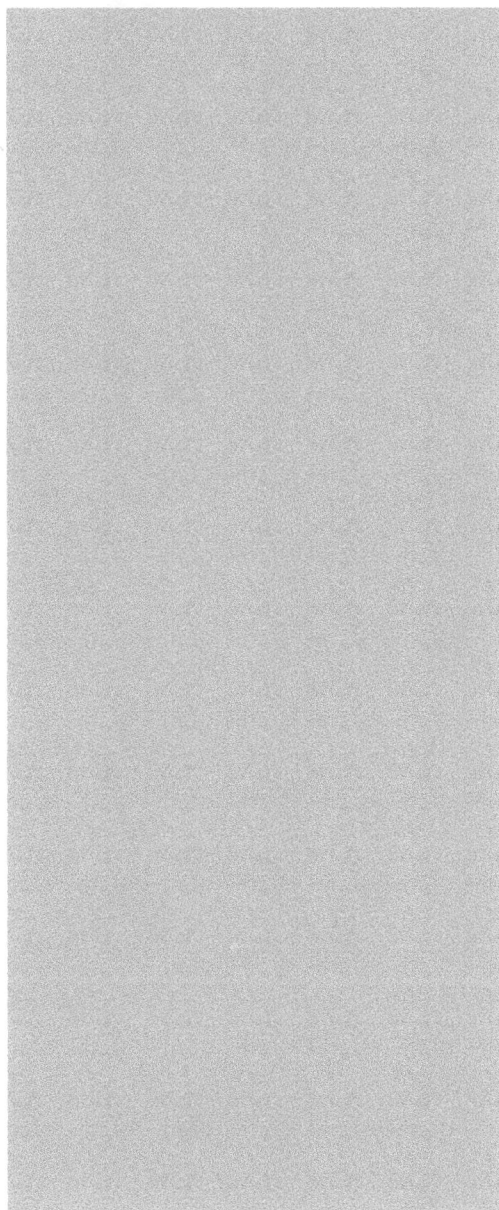

The findings of the Review Panel (RP) with respect to the review criteria are as follows:

Criterion 1

Are the design and execution of the spent nuclear fuel (SNF) canister consistent with established scientific and engineering principles and standards? In particular, is the Project Team (PT) aware of relevant scientific and engineering data and publications on remote welding for canisters?

Finding 1 of the RP

The design and execution of the SNF canister welding concept are consistent with established scientific and engineering principles and standards. The PT appears to be aware of relevant scientific and engineering data on remote welding for canisters, but did not cite specific publications and references.

As presented by the PT, the U.S. Department of Energy (DOE) has developed a design concept for a set of standardized canisters for the handling; interim storage; transportation; and disposal in the national repository of DOE SNF. The standardized DOE SNF canister must be capable of handling the DOE SNF in a variety of potential storage and transportation systems. It must also be acceptable to the repository, based on current and anticipated future requirements. This expected usage mandates a robust design.

The material selection for the DOE SNF canisters was made taking into account the following factors:

1. The most demanding load on the canisters is expected to be the accidental drop, so a strong and ductile material was needed.
2. The desirability to avoid the problems associated with inter-granular stress corrosion cracking (IGSCC) makes a low-carbon material (base metal and welds) a requirement.

3. The canisters could be in a moist environment, so a material with high-corrosion resistance is a priority.
4. Exotic materials are more expensive than commonly-available materials. Many of these canisters could be manufactured using common materials provided they meet the structural and ASME (2001) code requirements. The use of common materials would result in significant cost savings.

It appeared that a 304L stainless steel would satisfy the material selection requirements listed above, and was thus chosen for the DOE SNF design concept canisters. Later, the DOE SNF canisters were modified to use 316L stainless steel instead of 304L since 316L has a better resistance to pitting corrosion than 304L.

The canister design has four unique geometries, with lengths of approximately 3 m (10 ft) or 4.5 m (15 ft), and an outside nominal diameter of 457 mm (18 inches) or 610 mm (24 inches). The canister has been developed to withstand a drop from 9 m (30 ft) onto a rigid (flat) surface, sustaining only minor damage (but no rupture) to the pressure (containment) boundary. The majority of the end drop-induced damage is confined to the skirt and lifting/stiffening ring components, which can be removed (if desired) after an accidental drop. The design concept for the DOE SNF canister includes the following:

1. Body made of seamless or longitudinally-welded pipe 457 mm (18-inch) diameter by 9.5 mm (3/8-inch) nominally thick, or 610 mm (24-inch) diameter by 12.7 mm (1/2-inch) thick, 316L stainless steel.
2. Heads are ASME-flanged and dished having 9.5 mm (3/8-inch) thickness for the 457 mm (18-inch) diameter canister, and 12.7 mm (1/2-inch) thickness for the 610 mm (24-inch) diameter canister, 316L stainless steel.
3. Skirts made of seamless or longitudinally-welded pipe (316L stainless steel) to match the body in diameter and thickness, which were 203 mm (8 inches) long for the 457 mm (18-inch) canister, and 228 mm (9 inches) long for the 610 mm (24-inch) canister.
4. Lifting rings made of plate (316L stainless steel), 25 mm (1-inch) wide by 12.7 mm (1/2-inch) thick, located just within the outer end of each skirt.

5. Interior impact plates made of 51 mm (2-inch) thick plate (A36 carbon steel) flat on one side for the contents to bear-on, and contoured on the other side to match the inside surface of the head.
6. Weight limit of 2,721 kg (6,000 lbs) for the 457 mm (18-inch) diameter canister, and 4,535 kg (10,000 lbs) for the 610 mm (24-inch) diameter canister.

A DOE SNF canister will be located within another secondary container by placing one end in the package and then sliding (or lowering) the canister into position. If the canister design included anything that protruded from the canister body, then installation (and removal) within the secondary container could be difficult. Therefore, it was desirable to make the canister body smooth (with no protrusions). This was easily achieved by making the canister body exterior cylindrical in shape. A cylindrical canister body exterior was easily produced using a standard pipe section. Using standard pipe sizes does not require: a special run at a mill (or the additional manufacturing efforts of rolling plates into a cylindrical shape); longitudinally seam-welding; and circumferentially welding sections to obtain the required length.

The PT did not address potential fractures of the canister final closure weld due to dynamic loads experienced during transportation and handling. The final closure weld is hereafter referred to as the seal weld. The PT did not assess the applicability of codes and standards dealing with dynamic loading.

The use of the proposed backing ring enhances the consistent production of full-penetration welds. However, the presence of the backing ring also increases the potential for cracking and/or corrosion in the welded zone. Moreover, the presence of the backing ring presents a potential source for ultrasonic testing artifacts. The PT did not address the significance of the backing ring design and its fit-up to ensure an acceptable weld.

The head and canister assembly geometric dimensioning and tolerances significantly affect the ability to make an acceptable weld. The PT did not address the details of the drawing number 507693 rev. 4 (DOE 1999) and the adequacy of these tolerances on the quality of the welds.

Criterion 2

Based on the extent of prior use, research, development, and application of the proposed welding process, has the PT demonstrated the appropriateness of the approach? In particular, has the PT demonstrated sufficient understanding of the baseline technology?

Finding 2 of the RP

Based on the extent of prior use, research, development, and application of the proposed welding process, the PT has demonstrated the appropriateness of the approach. In general, the PT has demonstrated sufficient understanding of the baseline technology.

The PT is tasked to develop a remote welding and examination process that can deposit and inspect the seal weld. The seal weld of the SNF-standardized canisters must be made after the DOE SNF is loaded into the canisters and the closure head is placed. As such, the welding, nondestructive examination, and repair will be performed remotely in a hot cell.

The objectives of this remote welding, nondestructive examination, and repair system are to:

1. obtain high integrity welds.
2. provide permanent and retrievable weld and nondestructive examination records meeting ASME (2001) Code requirements.
3. minimize the radiation exposure to "As Low As Reasonably Achievable."
4. minimize the waste streams generated during the welding, testing, and repair processes.

Key issues regarding the welding and nondestructive examination of the standardized DOE SNF canisters final seal welds include:

1. The ability to operate, inspect, and detect remotely in a high-radiation environment and at a relatively high temperature in accordance with ASME (2001) Code Section III, Division 3 requirements
2. Resultant weld quality and documentation

3. Long-term thermal/micro-structural stability of materials
4. Storage environment
5. Final weld joint design

The weld inspection equipment is stationed on a carrousel that rotates around the SNF-standardized canister. The weld power supply and inspection electronics are housed behind a shielding device to minimize the equipment exposure to a high-radiation environment. Each of the three towers of the carrousel supports a subsystem (welding, inspection, or repair). Only the end effector portion of the equipment is exposed to a high-radiation environment. All repair cavities will have a consistent geometry. The Eddy Current Testing (ET) probe will be shaped to fit the groove geometry. The ultrasonic testing (UT) subsystem is envisioned to consist of a phased array piezoelectric probe. Iterative "grind and inspect" steps will be performed to verify the removal of a defect.

The development plan is directed towards developing welding, nondestructive examination, and analysis techniques (including equipment) that can not only be used iteratively to make and inspect the final closure seal welds on the standardized DOE SNF canisters, but also addresses the aforementioned key issues. The PT performed vendor assessments for the components of the welding, inspection, and repair system. The requirements listed for each component include general requirements as well as environmental requirements. The environmental requirements are based on assumptions on the fuel to be contained and expected temperature.

The specific component requirements were well defined. However, the PT did not present the requirements for the relationships among the different components; between the system and the hot cell; and between the system and the operator. For example, the PT did not sufficiently discuss a conceptual plan for automatic motion control; calibration; tool change out; and coordination of multiple axes positioning equipment in the execution of welding, inspection, maintenance, and repair operations. The gross positioning control and its coupling to the welding parameters (such as voltage, current, and wire speed) are critical to the success of this project. Providing a welding schedule overview (coupled with a description

131

of its implication on the equipment requirements); representative weld transverse cross-sections that reveal the weld passes; and micro-structure (including ferrite content) is important. In addition, there was no discussion of features to address safety issues such as emergency stops.

Conventional welding and inspection methods cannot be applied in this case due to the harsh environment and the lack of shielding. In the USNRC licensing strategy for the Idaho National Engineering and Environmental Laboratory SNF Canister, helium leak-testing, remote welding, and real time nondestructive examination (NDE) are important. In addressing these issues, qualified data must be generated to provide reasonable assurance that the fabrication process will produce a canister with a high-quality seal weld.

The end result is expected to be the development of an integrated welding, inspection, and repair system that will meet the ASME (2001) Boiler and Pressure Vessel Code, Section III, Division 3. Welding procedures, NDE procedures, and personnel must be qualified to the requirements set forth in this code. Joint designs that minimize welding defects; minimize weld process heat input; and facilitate inspection (especially volumetric) are required and will be developed using an integrated approach. The ASME (2001) Section III, Division 3 requires all final welds to be inspected using visual, surface, and volumetric examination techniques. Remotely-operated equipment and techniques must be developed to perform closure weld inspections in the hot cell environment.

The PT provided a description of a suggested set of NDE testing procedures, but did not evaluate their performance in a harsh environment. Both the proposed eddy current and phased array ultrasonic testing techniques need to be evaluated under operating conditions for signal/noise ratio, repeatability, and durability (transducers and cabling). The eddy current probe needs to be evaluated for wear. Moreover, it is necessary to assess whether probe wobble, mechanical compliance, or liftoff (due to weld roughness or imprecise alignment) are problems. The ultrasonic phased array probe needs to be evaluated for:

1. loss of signal due to water couplant problems and/or possible damage to the piezoelectric properties
2. ability to scan the entire weld walls as well as the weld volume

Other possibilities, such as non-contact ultrasonic techniques and laser-profiling systems, were discussed in a limited way. These techniques may be helpful in case the water couplant problem for the ultrasonic technique is insurmountable due to temperature problems.

Although the PT has considered validation procedures for their NDE testing, suggestions for the content of a comprehensive validation plan were not presented. The validation must include:

1. representative weld samples prepared with a statistically-significant number of well-characterized defects (ASME [2001] Boiler and Pressure Vessel Code, Section III, Division 3)
2. a scan plan designed to improve inspectability; maximize the probability of flaw detection; and minimize the number of false calls

The initial focus was shown for components and process tools. Relatively little was presented regarding the integration of these components into an overall conceptual design which includes interfaces (among the different components; between the system and the hot cell; and between the system and the operator); system architecture; and performance requirements. Moreover, for human operations external to the hot cell, it is necessary to address ergonomic and interface issues.

Criterion 3

Have environmental—including human health—risks been adequately identified and addressed for the conceptual design?

Finding 3 of the RP

The PT has adequately identified and addressed the environmental—including human health—risks for the purposes of a conceptual design.

The approach presented by the PT would significantly reduce personnel radiation exposure; reduce secondary waste; and improve productivity. The integrated system for automated welding would significantly reduce personnel exposure during operation by providing one complete system and one procedure. The NDE technology available from commercial vendors is for post weld testing that requires additional setup and testing procedures independent of the welding process. In addition to the staging, setup, and removal of the automated welding equipment, technicians and quality inspectors are required to setup and then remove the NDE equipment. These extra steps can be eliminated, thus significantly reducing the exposure time with the integration of the in situ NDE technology. Additional exposure associated with post weld repairs would be reduced by providing notification of weld defects on partially-completed welds, thus reducing the amount of weld filler material that is normally removed by grinding on completed welds.

The PT discussed best practices in hot cell design to minimize environmental/human health risks. The minimization of the secondary waste streams was adequately addressed. For example, the secondary waste stream produced through grinding is minimized by performing pass-by-pass inspection and producing a smooth final cap pass to enable direct inspection. This implies reduced airborne contamination and less time associated with weld repair.

A principal assumption made by the PT to meet the required exposure levels for the purposes of human operator entry was the removal (from the hot cell) of the canister coupled with the shielding of the portal for the canister (to the hot cell). However, the PT has not yet estimated the residual contamination level, and the frequency and time required for human involvement in operations, maintenance, and repair internal to the hot cell, which are necessary for the final design of the chosen concept. The placement of the electronics inside the hot cell and the reliance on wireless communications could be problematic.

Criterion 4

Has the PT demonstrated that the proposed welding process meets or exceeds customer requirements?

Finding 4 of the RP

The concept of the welding, inspection, and repair system does have the potential to meet or exceed customer requirements. The PT recognized that it has developed only a conceptual design, and it is aware of the major hurdles associated with the high-radiation field and relatively high temperatures.

To mitigate the risk of premature component failures, the conceptual design includes the replacement of vulnerable components with hardened equipment. The PT recognized that other options might need to be explored for ultrasonic NDE due to the relatively high-operation temperature.

The PT has recognized that there are conflicting requirements that are outside its purview. For example, not requiring the canister to rotate has added unnecessary complexity to the concept. In addition, the PT may need to request and gain approval of variances to the ASME (2001) Boiler and Pressure Vessel Code, Section III, Division 3.

Criterion 5

Has the PT demonstrated that the proposed welding process is optimized? In particular, has the PT provided sufficient evidence indicating that a less-automated solution cannot meet customer requirements?

Finding 5 of the RP

The PT has performed only a conceptual design. It is premature to consider optimization of a conceptual design. As a first step to optimize the system, it is necessary to define a set of system-level performance requirements to meet the customer functional requirements.

The PT provided evidence indicating that, subject to the constraints imposed by the customer, a less-automated solution would not meet the specified requirements.

Criterion 6

Based on the technical merit of this project, is its likelihood of successful implementation reasonably high? Should this project be continued?

Finding 6 of the RP

Based on its technical merit, there is a reasonably high likelihood for the successful implementation of this project. This project should be refined and continued in the development of a detailed design.

Additional Finding of the RP: Finding 7

The PT has demonstrated good design principles in its approach and awareness of the problems related to the canister welding concept. The PT is capable to address the hurdles encountered in the implementation of this conceptual design.

Additional Finding of the RP: Finding 8

There is a discrepancy between:

1. the specification for the primary service temperature (350°F) for a DOE SNF canister when it is not inside another container (DOE 1999) and the specification for the maximum operating temperature (300°F) for a DOE SNF canister when it is not inside another container (DOE 1999); and
2. the temperature (200°F) assumed by the PT during its oral presentation of the conceptual design.

This discrepancy must be resolved. Certain components of the conceptual design of the welding, inspection, and repair system are expected to fail at temperatures higher than 200°F.

Additional Finding of the RP: Finding 9

This project cannot be performed in isolation, independent of other activities. Successful completion is dependent upon the effective coordination among the different groups involved in the formal design of the:

1. welding, inspection, and repair system
2. hot cell facility
3. SNF canister
4. packaging and transportation platform

RECOMMENDATIONS

Based on a careful assessment of the information presented to the Review Panel (RP) and the findings developed in response to the review criteria, the RP provides the following recommendations:

1. This project should be continued provided the remainder of these recommendations is seriously considered.
2. An organizational structure should be created in order to ensure an appropriate coordination among the different groups involved in the formal design of the welding, inspection, and repair system; hot cell facility; spent nuclear fuel (SNF) canister; and packaging and transportation platform.
3. The Project Team (PT) must resolve the discrepancy between the temperatures provided in the specification for the DOE SNF canister and the temperature assumed in the conceptual design.
4. The PT must define geometric dimensioning and tolerances (GD&T) for the head and canister assembly to ensure acceptable quality welds.
5. The PT must investigate the potential fractures of the canister seal weld due to dynamic loads experienced during transportation and handling. The applicability of codes and standards dealing with dynamic loading should be assessed.
6. The PT should evaluate alternative designs to the backing ring that do not increase the potential for cracking and/or corrosion in the welded zone.
7. The proposed nondestructive examination (NDE) eddy current and ultrasonic techniques should be tested and evaluated under operating conditions of radiation and temperature to determine the feasibility of the proposed approaches.
8. The PT should evaluate in greater detail non-contact ultrasonic techniques such as laser ultrasonics; electromagnetic acoustic transducers (EMATs); and air-coupled ultrasonics for the inspection process.
9. The PT should develop a validation plan sufficient to qualify the selected NDE techniques for operation in a harsh environment with high reliability.
10. The PT should develop a formal specification that includes performance requirements; system architecture; and interfaces (among the

different components; between the system and the hot cell; and between the system and the operator).

11. The PT should estimate the residual contamination level, and the frequency and time required for human involvement in operations, maintenance, and repair internal to the hot cell.

12. In a subsequent design, the PT should develop a plan for automatic motion control; calibration; tool changeout; and coordination of multiple axes positioning equipment in the execution of welding, inspection, maintenance, and repair operations.

References

ANSI (American National Standards Institute). Design criteria for an independent spent fuel storage installation (dry type), ANSI/ANS 57.9. New York, NY: American National Standards Institute; 1992.

ASME (American Society of Mechanical Engineers). Boiler and pressure vessel code. New York, NY: American Society of Mechanical Engineers; 2001.

ASME (The American Society of Mechanical Engineers). Manual for peer review. Washington, DC: ASME; 2003.

ASME/RSI (The American Society of Mechanical Engineers/Institute for Regulatory Science). Assessment of technologies supported by the U.S. Department of Energy; results of the peer review for fiscal year 1997, CRTD Vol. 47. New York: ASME; 1997.

ASME/RSI (The American Society of Mechanical Engineers/Institute for Regulatory Science). Assessment of technologies supported by the U.S. Department of Energy; results of the peer review for fiscal year 1998, CRTD Vol. 50. New York: ASME; 1998.

ASME/RSI (The American Society of Mechanical Engineers/Institute for Regulatory Science). Assessment of technologies supported by the U.S. Department of Energy; results of the peer review for fiscal year 1999, CRTD Vol. 56. New York: ASME; 1999.

ASME/RSI (The American Society of Mechanical Engineers/Institute for Regulatory Science). Assessment of technologies supported by the U.S. Department of Energy; results of the peer review for fiscal year 2000, CRTD Vol. 61. New York: ASME; 2000.

ASME/RSI (The American Society of Mechanical Engineers/Institute for Regulatory Science). Strategy for remediation of groundwater contamination at the Nevada Test Site; technical peer review report; report of the review panel, CRTD Vol. 62. New York: ASME; 2001a.

ASME/RSI (The American Society of Mechanical Engineers/Institute for Regulatory Science). Requirements for disposal of remote-handled transuranic wastes at the waste isolation pilot plant; technical peer review report; report of the review panel, CRTD Vol. 63. New York: ASME; 2001b.

ASME/RSI (The American Society of Mechanical Engineers/Institute for Regulatory Science). Assessment of technologies supported by the U.S. Department of Energy; results of the peer review for fiscal year 2001, CRTD Vol. 64. New York: ASME; 2001c.

ASME/RSI (The American Society of Mechanical Engineers/Institute for Regulatory Science). Waste isolation pilot plant initial report for poly-chlorinated biphenyl disposal authorization; technical peer review report; report of the review panel, CRTD Vol. 65. New York: ASME; 2002a.

ASME/RSI (The American Society of Mechanical Engineers/Institute for Regulatory Science). Airborne release fractions; technical peer review report; report of the review panel, CRTD Vol. 68. New York: ASME; 2002b.

ASME/RSI (The American Society of Mechanical Engineers/Institute for Regulatory Science). The beryllium oxide manufacturing process; technical peer review report; report of the review panel, CRTD Vol. 69. New York: ASME; 2002c.

ASME/RSI (The American Society of Mechanical Engineers/Institute for Regulatory Science). Assessment of technologies supported by the U.S. Department of Energy; results of the peer review for fiscal year 2002; CRTD Vol. 70. New York: ASME; 2002d.

ASME/RSI (The American Society of Mechanical Engineers/Institute for Regulatory Science). Radionuclide transport in the environment; technical peer review report; report of the review panel; CRTD Vol. 71. New York: ASME; 2002e.

ASME/RSI (The American Society of Mechanical Engineers/Institute for Regulatory Science). Review of selected nuclear safety programs at Savannah River Site; technical peer review report; report of the review panel, CRTD Vol. 72. New York: ASME; 2003a.

ASME/RSI (The American Society of Mechanical Engineers/Institute for Regulatory Science). Hanford Site 100 B/C Risk Assessment Pilot Project; technical peer review report; report of the review panel, CRTD Vol. 73. New York: ASME; 2003b.

ASME/RSI (The American Society of Mechanical Engineers/Institute for Regulatory Science). Salt waste processing facility technology readiness, CRTD Vol. 75. New York: ASME; 2003c.

DOE (U.S. Department of Energy). Preliminary design specification for Department of Energy spent nuclear fuel canisters, DOE/SNF/REP-011. Rev. 3. Volume I and II. Idaho Falls, ID: Idaho National Engineering and Environmental Laboratory; (August 17) 1999.

DOE (U.S. Department of Energy). Standard contract for disposal of spent nuclear fuel and/or high level radioactive waste, 10 CFR 961, 2003. Available at http://www.access.gpo.gov/nara/cfr/

McJunkin, T. Spent nuclear fuel standardized canister closure concept. Idaho Falls, ID: Idaho National Engineering and Environmental Laboratory; (November 3) 2003.

NWPA (Nuclear Waste Policy Act). Public law 97-425; 1982.

NWPAA (Nuclear Waste Policy Amendments Act). Public law 100–203; 1987.

Presidential Memorandum. Disposal of defense waste in a commercially repository. (April 30) 1985.

RSI (Institute for Regulatory Science). Handbook of peer review. Columbia, MD: RSI; 2003.

USNRC (U.S. Nuclear Regulatory Commission). Disposal of high level radioactive wastes in geologic repositories, 10 CFR 60, 2003a. Available at http://www.access.gpo.gov/nara/cfr/

USNRC (U.S. Nuclear Regulatory Commission). Packaging and transportation of radioactive material, 10 CFR 71, 2003b. Available at http://www.access.gpo.gov/nara/cfr/

USNRC (U.S. Nuclear Regulatory Commission). Licensing requirements for the independent storage of spent nuclear fuel, high level radioactive waste, and reactor related greater than class C waste, 10 CFR 72, 2003c. Available at http://www.access.gpo.gov/nara/cfr/

Watkins, A.D. Vendor assessments weld, inspection and repair system. Idaho Falls, ID: Idaho National Engineering and Environmental Laboratory; (November 3) 2003a.

Watkins, A.D. SNF welding and inspection system compliance & verification matrix. Idaho Falls, ID: Idaho National Engineering and Environmental Laboratory; (November 3) 2003b.

Biographical
Summaries

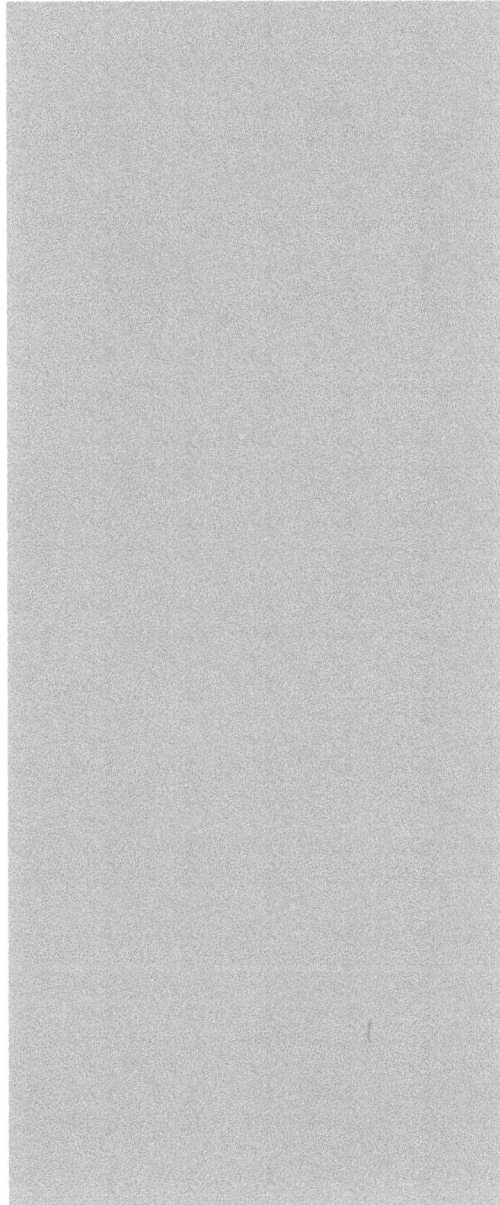

Gary A. Benda is President of U.S. Energy Corp.—an environmental management firm specializing in radioactive mixed waste management, health physics, decontamination and decommissioning, and technology development. Previously, he was Vice-President, General Manager of the Programs Division for NUKEM Nuclear Technologies, Inc. His responsibilities included developing and maintaining federal programs in North America that specialized in engineering and waste-processing services. Prior to NUKEM, he spent over 17 years with Chem-Nuclear Systems/WMX Technologies in various management roles. He also directed the site investigation, geophysical analysis, site screening, and license application, as well as managed the public hearings and licensing operations associated with local and national regulatory agencies for new low-level waste sites. He has over 20 years of experience in environmental restoration, technology development, and waste management, and has instructed over 20 national and international professional courses on radioactive waste management, mixed waste, and technology development. He is a member of the American Society of Mechanical Engineers (ASME), American Nuclear Society, and Health Physics Society. He has served as Chair of the ASME National Mixed Waste Committee, Environmental Remediation Committee, and Environmental Engineering Division. He has also chaired over 100 technical sessions at numerous national and international conferences on environmental management. He has authored and coauthored various scientific papers, reports, book chapters, and articles on the nuclear environment. Gary Benda is a Certified Health Physicist. He received a B.S. in Health Physics from Oklahoma State University, an M.S. in Applied Nuclear Science from Georgia Institute of Technology, and an M.B.A. from Seattle City University.

George E. Cook is currently Professor of Electrical Engineering, and Associate Dean for Research and Graduate Studies, Engineering School, Vanderbilt University, Nashville, TN, His studies focus on various aspects of robotic welding, including modeling and control of welding processes. He has also studied statistical process control applications to welding processes and Jacobian control for space manipulators. He started his career as an instructor; became an Assistant Professor; was promoted to Associate Professor; and eventually full professor, serving in various leading positions at Vanderbilt. He is a fellow of the American Welding Society; a fellow of the Institute of Electrical and Electronics Engineers; and has served as member of the Board of the American Welding Institute. He was also associated with a number of private corporations including CRC Welding Systems; Merrick Corporation; Advanced Control Engineering; and Industrial Electronics Laboratory. He has been awarded over 30 patents in the U.S. and foreign countries, and has over 200 publications. He has received numerous awards, including the James F. Lincoln Arc Welding Foundation Gold Award for the development of an adaptive robotic arc welding control methodology. George E. Cook received a B.E. degree from Vanderbilt University; an M.S. degree from the University of Tennessee; and a Ph.D. from Vanderbilt University, all in Electrical Engineering. He is a registered professional engineer in Tennessee, Kentucky, Alabama, and Wisconsin.

Erich W. Bretthauer is a consultant. Previously, he held the position of research professor at the University of Nevada-Las Vegas from January 1993 to March 1995. In that capacity, he served as Executive Director of Nevada Industry, Science Engineering & Technology, a public-private partnership which developed programs to enhance the scientific infrastructure of the state of Nevada. He was also the Assistant Administrator for Research and Development at the U.S. Environmental Protection Agency (EPA) from March 1990 until January 1993. In that capacity, he managed the Research and Development activities of a large and multi-disciplinary agency. Erich Bretthauer rose through the ranks of the EPA and served in a number of capacities ranging from a bench scientist to policy manager at national and international levels. He directed the EPA's emergency and long-term monitoring program after the accident at Three Mile Island, as well as its bioremediation program in Prince William Sound after the Valdez oil spill. Erich Bretthauer was the leader of the U.S. delegation and co-leader of a five-year North Atlantic Treaty Organization project which focused on exposures, risks, and measures to control dioxins. He also directed the EPA's ecological research program, and was Director of EPA's Environmental Monitoring Systems Laboratory in Las Vegas. He is a member of Sigma Xi; the American Chemical Society; and the American Water Works Association; and has served on the Federal Advisory Committee to the Civil Engineering Research Foundation. Erich Bretthauer is the author and coauthor of numerous papers, reports, and other publications. He received his B.S. and M.S. in chemistry from the University of Nevada, Reno, NV.

Ernest L. Daman is Chairman Emeritus of Foster Wheeler Development Corporation where he previously served as Director of Research and Chairman of the Board. He also held the position of Senior Vice President at the parent company, FWC. He is a Past President of American Society of Mechanical Engineers and was elected to the National Academy of Engineering. Ernest Daman is a Fellow of the Institute of Energy (England) and the American Association for the Advancement of Science, and Past Chairman of the American Association of Engineering Societies. He served on several American Society of Mechanical Engineers committees as member or chairman. Ernest Daman is the author of numerous papers and holds 18 patents. He was responsible for the design and development of a combined steam gas turbine plant, fluidized bed combustion, fast breeder reactor components, supercritical steam generators, environmental control processes, and advanced high-efficiency power generation systems. Ernest Daman received his B.M.E. degree from the Polytechnic Institute of Brooklyn.

Irwin Feller is Director of the Institute for Policy Research and Evaluation and Professor of Economics at The Pennsylvania State University, where he has been on the faculty since 1963. His current research interests include the economics of academic research, the University's role in technology-based economic development, and the evaluation of federal and state technology programs. He is the author of *Universities and State Governments: A Study in Policy Analysis*, and over 100 refereed journal articles, final research reports, book chapters, reviews, and numerous papers presented to academic, professional, and private organizations. He is former Chair of the Committee on Science, Engineering, and Public Policy, American Association for the Advancement of Science. Irwin Feller was the American Society of Mechanical Engineers Pennsylvania State Fellow in 1996–1997. He has been appointed to the National Research Council's Committee on Science, Engineering, and Public Policy; International Benchmarking of U.S. International Competitiveness-Immunology; Transportation Research Board, Research and

Technology Coordinating Committee, National Research Council; and National Institute of Standards and Technology-Manufacturing Extension Partnership National Advisory Board. Irwin Feller is Chair of the National Science Foundation's Advisory Committee on Social, Behavioral, and Economic Sciences. He received a B.B.A. in Economics from the City University of New York and a Ph.D. in Economics from the University of Minnesota.

Robert A. Fjeld is Dempsey Professor of Environmental Engineering and Science at Clemson University. He coordinates the Department's nuclear environmental focus area, which is concerned with the environmental aspects of nuclear technologies including health physics, radioactive waste management, and risk assessment. Previously, he served as a faculty member in the Nuclear Engineering Department at Texas A&M University. He has active research on actinide transport in soils, instrumentation for measuring radioactivity in environmental samples, and environmental risk assessment. Robert Field is a member of the Health Physics Society, American Nuclear Society, Society for Risk Analysis, and the American Society of Mechanical Engineers, where he serves as newsletter editor for the Mixed Waste Committee. He has served on two NRC Committees studying decontamination and decommissioning issues. Robert Fjeld is the author or coauthor of over 80 technical publications and presentations on topics such as radiation measurements, environmental transport of radionuclides, risk assessment, and aerosol physics. Robert Fjeld received a B.S. degree from North Carolina State University; and an M.S. degree and a Ph.D. from The Pennsylvania State University, all in Nuclear Engineering. He is a registered Professional Engineer.

William T. Gregory, III is currently Principal of Vinculum Marketing Solutions. Prior to forming Vinculum, he was Director of Government Programs for Foster Wheeler Environmental Corporation, an engineering and construction firm providing environmental and waste management services to government and private sector clients worldwide. Previously, he held a number of operational and business development positions at equipment manufacturing and service provision firms supporting nuclear utilities, industrial and process industries, and government agencies.

His work has involved the management, processing, and disposition of hazardous, radioactive, and mixed wastes. He has also worked on the decontamination and decommissioning of nuclear facilities and on providing a wide range of environmental services in response to regulatory drivers. Prior to entering the private sector, he served with the U.S. Navy on nuclear submarines and at the operational command center for submarine operations in the Atlantic Fleet. William Gregory is actively involved with a number of international, national, and local organizations including the American Society of Mechanical Engineers and the American Nuclear Society. He is a founding member of the Board of Directors for the annual international Waste Management Symposium. William Gregory has served as an elected officer of several American Society of Mechanical Engineers divisions. He received a B.S. degree in Geology from the University of New Mexico, and an M.B.A. degree from Lamar University. He also attended naval nuclear power, nuclear weapons, and engineering schools as a U.S. Navy officer.

Nathan H. Hurt is currently an Executive Consultant working for Sharp and Associates and JAI Corporation. He has over 55 years of extensive experience in the nuclear and chemical fields. In the nuclear field he has experience in uranium enrichment engineering and production, design and construction for U.S. Department of Energy (DOE) weapons facilities and is currently involved in the decommissioning (decontamination and dismantlement) of nuclear facilities. His experience in the chemical industry is in the engineering and management of rubber chemicals, petrochemicals, and thermoplastics facilities. Previously he was Vice President and Director of Oak Ridge Operations for IDM Environmental Corporation. During the time he worked for IDM, he served as Corporate Sponsor or Program Manager for a multiple number of decommissioning projects at DOE facilities in Oak Ridge, TN; Pinellas, FL; the Savannah River Site (SRS); and several industrial chemical plants. Prior to that he was Director of Marketing and Projects for Los Alamos Technical Associates. During this time he was responsible for a multiple number of engineering studies and designs for DOE facilities at Rocky Flats, Hanford, Oak Ridge, SRS, and Portsmouth, OH. He previously worked for The Goodyear Tire and Rubber Company, managing the design and construction and operation of chemical facilities (synthetic rubbers and latices; vinyl monomer and polymer; and polyesters) both domestic and overseas. He served in various capacities at Goodyear Atomic Corporation, a subsidiary of Goodyear Tire, including Manager of Engineering, Deputy General Manager and President. Goodyear Atomic was the contractor for DOE for the operation of the Portsmouth Area Uranium Enrichment Facility. He is a Past President of the American Society of Mechanical Engineers (ASME). He has been a member of: the American Association of Engineering Societies' Board of Governors; the American Institute of Chemical Engineers; the American Society of Engineering Management; and the Institute of Nuclear Materials Management. He previously served on The Nuclear Engineering Advisory Board of Worcester Polytechnic Institute. Nathan Hurt received a B.S. degree in Mechanical Engineering from the University of Colorado and has done graduate work at Pennsylvania State University. He is a registered Professional Engineer in Ohio.

Adam S. Jacoff is currently a Project Leader in the Intelligent Systems Division of the National Institute of Standards and Technology (NIST) in Gaithersburg, MD. He has fifteen years experience designing and developing innovative robots for applications in large scale manufacturing, nuclear/hazardous waste handling, and machine vision, and holds three patents related to parallel kinematic robotic cranes. He was Test Director for the Army's first technology readiness level assessment of their experimental unmanned vehicle (XUV) which captured performance data over 700 missions in arid, vegetated and urban terrain totaling more than 500 km of autonomous travel. He recently designed and fabricated NIST's *Reference Test Arenas for Autonomous Mobile Robots* which isolate and test mobile robotic capabilities such as sensory perception, knowledge representation, planning, collaboration, and other autonomous behaviors. He uses these transportable arenas to objectively evaluate robot performance within repeatable testing environments representative of urban search and rescue scenarios in an effort to define and quantify performance metrics for autonomous mobile robots. These arenas have been replicated in the U.S. and internationally to host robotic search and rescue competitions. Mr. Jacoff has authored over twenty publications related to his research, including an award for best paper in the 2002 volume of Industrial Robot Journal. He has been awarded the U.S. Department of

Commerce Bronze Medal and Japan's Society of Instrumentation and Control Engineers (SICE) International Award. He holds a B.S. degree in Mechanical Engineering from the University of Maryland, and an M.S. degree in Computer Science from Johns Hopkins University.

Michael C. (M.C.) Kirkland is Vice President for the Southeastern Region of the Institute for Regulatory Science (RSI). In that capacity he leads various RSI projects related to the RSI mission in the southeastern U.S. Previously he was an independent consultant involved in peer review and various independent studies. For example, he led a team that performed an External Independent Review of the $1.3 billion Spallation Neutron Source Project at Oak Ridge, TN. He assisted in the planning and review of a management assessment at a U.S. Department of Energy (DOE) Site that involved the restart of a plutonium facility. He participated in planning, procurement, and review activities in the environmental remediation area that included decommissioning activities at a shut down nuclear test reactor; and designed and installed a ground water cleanup technology. M.C. Kirkland managed several environmental and construction projects that employed many soil investigative techniques including significant work with cone penetrometers. Additionally, he provided consulting services to a large environmental remediation services company regarding Dense Non-Aqueous Phase Liquid locating and removal techniques. During his tenure at the Savannah River Site (SRS) of DOE, M.C. Kirkland was a Technical Advisor, Project Manager, and Director of the Project Engineering Division. He evaluated nuclear and mixed waste conditions and aspects of high level wastes and spent nuclear fuel; determined material inventories; performed pollution prevention and environmental health and safety evaluations for a proposed waste treatment facility; served as technical advisor to a study administered by the Savannah River Operations Office; and developed integrated schedules defined for this project. M.C. Kirkland was director of the Project Engineering Division and managed the SRS design and construction program. He has been involved with waste management and environmental projects; cutting-edge technology programs; and worked with lasers and magnetic containment. He served as Director of the Waste and Fuel Cycle Technology Office, and planned and coordinated the programs of the DOE National High Level

Waste Technology Office; the SR Fuel Cycle Technology Program; and the Commercial Interim Spent Fuel Management Program. M.C. Kirkland holds a B.S. in Mechanical Engineering from the University of South Carolina. He is registered as a Professional Engineer in South Carolina.

Sindo Kou is currently Professor and Chair, Department of Materials Science and Engineering, University of Wisconsin, Madison, WI, having joined the University as Associate Professor. His studies focus on materials processing, which include welding; crystal growth; and casting and heat treating. He is also interested in transport phenomena including the roles of fluid flow; heat transfer and mass transfer in welding, crystal growth, casting and heat treating. Previously, he was an Associate and Assistant Professor at Carnegie-Mellon University, Pittsburgh, PA. He was also Associate Senior Research Engineer, General Motors Research Laboratory, Warren, MI; and Research Associate, University of Illinois, Urbana-Champaign, IL. He is a fellow of the American Welding Society, and a fellow of ASM International. Sindo Kou has published over 100 papers, and has authored several books including *Welding Metallurgy,* which has appeared in two successive editions. He is the recipient of numerous awards and has been awarded four patents. Sindo Kou received a B.S. degree in Chemical Engineering from National Taiwan University, Taipei, Taiwan; an M.S. degree in Materials Engineering from the University of Wisconsin, Milwaukee, WI; and a Ph.D. in Metallurgy from Massachusetts Institute of Technology, Cambridge, MA.

Peter B. Lederman is a consultant with over 48 years of experience in all facets of process engineering, environmental management, control, and policy development. This includes hazardous substance management; environmental remediation; environmental audit; pollution prevention; development of air pollution control devices; and reuse of waste products. He recently retired as Executive Director of the Center for Environmental Engineering & Science, Executive Director for Patents and Licensing, and Research Professor of Chemical Engineering and Environmental Policy at the New Jersey Institute of Technology. Peter Lederman managed major programs in industrial waste treatment research and development, and in oil and hazardous material spill control and remediation. Most recently, he was responsible for a study of the Economic Impact of Environmental Regulations. He has been responsible for technology transfer efforts including the maturing and licensing of innovative environmental technologies. He is a Fellow of the American Institute of Chemical Engineers; a Diplomat of the American Academy of Environmental Engineers; and a member of the ASME. He has served on several committees of the NRC and is the chair of the NRC Committee on Review and Evaluation of the Army Chemical Stockpile Disposal Program. He chaired the American Institute of Chemical Engineer's Environmental Division and is currently chair of its Societal Impacts Operating Council. Peter Lederman received a B.S.E., M.S.E., and Ph.D. (all in Chemical Engineering) from the University of Michigan in Ann Arbor, MI and is a registered Professional Engineer.

Betty R. Love is currently Executive Vice President of the Institute for Regulatory Science. In that capacity, she is responsible for the management of day-to-day operations of the Institute, and for administration of several projects. She is the Administrative Manager of a large-scale peer review program in collaboration with the American Society of Mechanical Engineers for a number of organizations including the U.S. Department of Energy. Her current research activities center around the development and implementation of a systematic approach to stakeholder participation, notably in scientific meetings. Previously, Betty Love was Director, Department of Training and Information within the Office of Environmental Health and Safety of Temple University in Philadelphia, PA. During that period she was instrumental in the development of a "Handbook of Environmental Health and Safety". She also developed and implemented a large-scale training program not only for the faculty and staff of the University but also for others. Betty Love is currently Managing Editor of *Technology*. She has published several papers in peer-reviewed journals; has edited a number of compendia; and is the primary author of *Manual for Public and Stakeholder Participation*. Betty Love received a B.S. degree in Business Administration from Virginia State University in Petersburg, VA, and an M.S. degree in Developmental Clinical Psychology from Antioch College in Yellow Springs, OH.

Jeffrey A. Marqusee is currently the Technical Director of the Strategic Environmental Research and Development Program (SERDP), and the Director of the Environmental Security Technology Certification Program (ESTCP). SERDP is a tri-agency (U.S. Department of Defense [DOD], U.S. Department of Energy, and U.S. Environmental Protection Agency) environmental research and development program managed by the DOD. SERDP supports research and development to solve environmental issues of relevance to DOD in the areas of cleanup, compliance, conservation and pollution prevention. ESTCP is a DOD-wide program designed to demonstrate innovative environmental technologies at DOD facilities. ESTCP provides for rigorous validation of the cost and performance of new environmental technologies in cooperation with the regulatory and end-user communities. Prior to his current position, Jeffrey Marqusee served as a program manager for environmental technology in the Office of the Deputy Under Secretary of Defense for Environmental Security. He was the principal advisor to the Deputy Under Secretary on environmental technology issues. Before joining DOD, he worked at the Institute for Defense Analyses, where he advised both DOD and National Aeronautics and Space Administration in the areas of remote sensing, environmental matters and military surveillance. Jeffrey Marqusee has worked at Stanford University, the University of California and the National Institute of Standards and Technology. He has a Ph.D. in Physical Chemistry from the Massachusetts Institute of Technology.

A. Alan Moghissi is currently President of the Institute for Regulatory Science (RSI), a non-profit organization dedicated to the idea that societal decisions must be based on best available scientific information. The activities of the Institute include research, scientific assessment, and science education at all levels—particularly the education of minorities. Previously, Alan Moghissi was Associate Vice President for Environmental Health and Safety at Temple University in Philadelphia, PA and Assistant Vice President for Environmental Health and Safety at the University of Maryland at Baltimore. In both positions, he established an environmental health and safety program and resolved a number of relevant existing problems in those institutions. As a charter member of the U.S. Environmental Protection Agency (EPA), he served in a number of capacities, including Director of the Bioenvironmental/ Radiological Research Division; Principal Science Advisor for Radiation and Hazardous Materials; and Manager of the Health and Environmental Risk Analysis Program. Alan Moghissi has been affiliated with a number of universities. He was a visiting professor at Georgia Tech and the University of Virginia, and was also affiliated with the University of Nevada and the Catholic University of America. Alan Moghissi's research has dealt with diverse subjects ranging from measurement of pollutants to biological effects of environmental agents. A major segment of his research has been on scientific information upon which laws, regulations, and judicial decisions are based—notably risk assessment. He has published nearly 400 papers, including several books. He is the Editor-in-Chief of *Technology: A Journal of Science Serving Legislative, Regulatory, and Judicial Systems*, which traces its roots to the *Journal of the Franklin Institute*—one of America's oldest continuously published journals of science and technology. Alan Moghissi is a member of the editorial board of several other scientific journals and is active in a number of civic, academic, and scientific organizations. He has served on a number of national and international committees and panels. He is a member of a number of professional societies including the American Society of Mechanical Engineers and is past chair of its Environmental Engineering Division. He is also an academic councilor of the Russian Academy of Engineering. Alan Moghissi received his education at the University of Zurich, Switzerland, and Technical University of Karlsruhe in Germany, where he received a doctorate degree in physical chemistry.

Lawrence C. Mohr, Jr., is currently Professor of Medicine, Biometry, and Epidemiology; and Director of the Environmental Biosciences Program at the Medical University of South Carolina. His areas of research and special interest include internal medicine and pulmonary disease—specifically diseases of the chest and respiratory system. An area of particular interest to Lawrence Mohr is environmental medicine, including molecular epidemiology and biomarker applications. He has been involved in studies related to environmental lung disease; pathophysiology; prevention and treatment of high altitude illness; high altitude physiology; risk assessment of environmental hazards and clinical epidemiology. Other areas of considerate interest to Lawrence Mohr are assessment of clinical outcomes; health policy analysis; and international health. This latter area includes: global epidemiology; medical relief operations; and health care in Central and Eastern Europe, as well as medical history—the impact of illness on world leaders. Previously, he held academic appointments as a Teaching Fellow in Medicine at the Uniformed Services University of the Health Sciences in Bethesda, MD. He was Associate Clinical Professor of Medicine and Emergency Medicine at George Washington University, Washington, DC. While in these institutions, he was a staff member of the Medical Support Group for the President of the United States. Lawrence Mohr was on the Medical Staff of Walter Reed Army Medical Center—where he completed his Internship and Residency in Internal Medicine—as well as George Washington University Hospital, both in Washington, DC. He has held Visiting Professorships at various universities. He served as Visiting Chief Resident at Presbyterian Hospital and Visiting Professor at the School of Nursing, both at Columbia University. Additionally, Lawrence Mohr was Visiting Professor of: William Beaumont Army Medical Center, Tulane University, University of Cincinnati, New York University, Brown University, East Carolina University, and the Mayo Clinic. Lawrence Mohr is a Fellow of the American College of Physicians and the American College of Chest Physicians. He is a member of several professional societies including: the American Federation for Medical Research; the Society for Risk Analysis; and the Wilderness Medical Society. Previously, he was on the Scientific Advisory Board for the Consortium in Environmental Risk Evaluation and the Savannah River Health Information System. He has authored or coauthored more than 60 articles, books, or technical publications. He received an A.B. degree in Chemistry as well as an M.D. degree, both from the University of North Carolina, Chapel Hill. Lawrence Mohr, Jr., is certified by the American Board of Internal Medicine.

Goetz K. Oertel's career in engineering, physics, chemistry, astronomy, and technical program management spans more than 40 years. He consults for industrial, academic, and governmental organizations in North and South America. As President and CEO of the Association of Universities for Research Astronomy, a nonprofit corporation, he engineered the initiation and completion of two 8-m aperture optical telescopes, and oversaw the Space Telescope Science Institute from before launch, through repair of the "Hubble flaw", to its successful operation. He initiated the conceptional phase of the Next Generation Space Telescope that will succeed Hubble as well as the Advanced Solar Telescope, and he oversaw the completion of ambitious ground-based astronomy facilities. He held technical and management positions in the U.S. Department of Energy, including Director of Defense Waste Management; Acting Manager of the Savannah River Operations Office; Deputy Manager of Albuquerque Operations Office; and Deputy Assistant Secretary for Safety, Health, and Quality Assurance. He had primary responsibility for the congressionally-mandated Defense Waste Management Plan, and for managing the related technology development, operations, and projects. He led the initiation of the Defense Waste Processing Facility, and saw it and the Waste Isolation Pilot Plant through technical, managerial, stakeholder, and political challenges. He was National Aeronautics and Space Administration Space Science Chief and Program Manager, and Aerospace Engineer at Langley. He was a Fellow in the White House with the President's Science Advisor and the Office of Management and Budget's Space and Energy branch. He chaired the Westinghouse West Valley Corporation Technical Advisory Group for high-level nuclear waste vitrification and management before, during, and after that project's successful vitrification campaign. He is a member of the American Physical Society, Sigma Xi, and other professional organizations. He is a Fellow of the American Association for the Advancement of Science. He is Chair or member of boards and committees of the National Research Council; George Mason University; the American Society of Mechanical Engineers; International University Exchange; and Westinghouse West Valley Corporation. He is a founding member of the Editorial Board for "Data Science", the new international on-line journal of Codata. He published numerous peer-reviewed papers and was awarded two patents. Trained as electrical engineer and physicist, he received a Vordiplom in Physics and Chemistry from the University of Kiel while on German industrial and governmental scholarships, and a Ph.D. in Physics from University of Maryland at College Park under a Fulbright scholarship.

Richard Rosenberg is currently a private consultant with experience in the design, development, manufacturing liaison, testing, field installation, and post installation and repair of special-purpose mechanical equipment. He worked for more than 25 years for General Atomic in a number of positions where he fostered the development of High Temperature Gas Cooled Reactor (HTGR) technology. He started as Senior Staff Engineer of the Fuel Handling Branch where he was responsible for the design of remotely operated fuel handling machines. He became Program Engineer of the Rotating Machinery Branch and the Fuel Handling Branch and oversaw the design, development, testing, field installation, maintenance, and manufacture of the Fort Saint Vrain (FSV) rotating machinery as well as the fuel handling equipment and control rod drives. Richard Rosenberg advanced to Manager of the Fuel Handling Branch; the Mechanical Equipment Branch; and later became Manager for Systems and Components for FSV services. As Manager, his responsibilities included all mechanical systems and components under the purview of General Atomic for the FSV reactor. He also served as Principal Engineer on the Reactor Development and Technology Staff for the U.S. Atomic Energy Commission. There he provided broad overview monitoring of the design, fabrication, and manufacturing efforts for the Fast Flux Test Facility sodium-cooled reactor-pressure vessel and guard vessel. As a Senior Engineer for Westinghouse he was responsible for the Shippingport PWR reactor refueling equipment and participated in the design of the reactor vessel. He was employed with the Oak Ridge National Laboratory where he designed special purpose remote handling equipment and special purpose chemical processing equipment. Richard Rosenberg held various positions as a volunteer with the American Society of Mechanical Engineers (ASME). He was elected president of ASME in 1987; and has been awarded Honorary Member status. He has served as a member and Chair of the Board of Governors; and as a member of the Board of Trustees of the ASME Foundation. He has been the recipient of the Dedicated Service Award and the Centennial Medallion as well as San Diego Engineer of the Year. He is affiliated with the Civilian Research and Development Foundation as a member of a team reviewing and recommending proposals for joint Russian and American research projects. He holds patents for a High Temperature, High Pressure Sampling System; a Rod Handling System;

and a Shielded Cask. Richard Rosenberg has a B.S. Degree in Mechanical Engineering from the University of Tennessee. He is a registered Mechanical Engineer in Pennsylvania; and prior to retiring registered Nuclear Engineer in California; and a Chartered Engineer in the United Kingdom.

Sorin R. Straja is currently Vice President for Science and Technology of the Institute for Regulatory Science. He has over 20 years of expertise in mathematical modeling and software development as applied in chemical engineering and risk assessment. Previously he served as Assistant Professor of Biostatistics with Temple University, Philadelphia; as Director of the Department of Occupational Health and Safety of Temple University, Philadelphia; and as a chemist with University of Maryland at Baltimore. Sorin Straja has extensive experience in the chemical industry where he worked as a senior R&D consultant with the Chemical and Biochemical Energetics Institute, and as a plant manager with Chemicals Enterprise Dudesti and Plastics Processing Bucharest from Romania. He was an Assistant/Adjunct Professor of Chemical Engineering with the Polytechnic Institute Bucharest. Sorin Straja is the author of two books and 44 scientific papers published in internationally recognized and peer-reviewed journals. He was an editor of *Environment International*, and currently is a contributing editor of *Technology*. Sorin Straja received a Certificate of Appreciation for Teaching from Temple University, the "Nicolae Teclu" Prize of the Romanian Academy, and a Certificate of Appreciation from U.S. Department of Agriculture for significant volunteer contributions. He is a Fellow of the Global Association of Risk Professionals, and a member of the American Chemical Society, American Institute of Chemical Engineers, Society for Risk Analysis, and New York Academy of Sciences. Sorin Straja holds a M.S. in Industrial Chemistry and a Ph.D. in Chemical Engineering both from Polytechnic Institute Bucharest.

Glenn W. Suter, II is currently Science Advisor at the National Center for Environmental Assessment of the U.S. Environmental Protection Agency (EPA) in Cincinnati, OH. Previous to his current position, he was at Oak Ridge National Laboratory, initially as Research Associate and gradually rising to Science Leader at the Environment Science

Division of the Laboratory. His interest has focused on Ecotoxicology in general and Ecological Risk Assessment in particular. He is one of the developers of the most widely-used methodology for Ecological Risk Assessment. This method has been applied to the impact of pollutants on fish, contaminated soils, production of synthetic fuels, and various other ecosystems. Glenn Suter has lectured widely, both nationally and internationally on Ecological Risk Assessment. He is currently a member of the U.S. EPA's Risk Assessment Forum. He has been a member of numerous panels and has consulted with various governmental agencies and private organizations, including the Council of Environmental Quality. He was a member of the Scientific Review Panel for Savannah River Ecology Laboratory; the National Science Foundation Panel on Decision Making and Valuation for Environmental Policy; and the U.S. EPA Science Advisory Board and Conservation Foundation, Ecosystem Valuation Forum. In addition, he was a member of the International Institute of Applied Systems Analysis Task Force on Risk and Policy Analysis and the Council on Environmental Quality. He was a member of the Board of Directors, for the Society for Environmental Toxicology and Chemistry. Glenn Suter is presently on the Editorial Board of *Environmental Health Perspectives* and *Human and Ecological Risk Assessment.* Previously, he was on the Editorial Board of *Handbook of Environmental Risk Assessment and Management* and *Environmental Toxicology and Chemistry.* Glenn Suter is the author of three books and is author and coauthor of over 200 publications. He received a B.S. degree in Biology from Virginia Polytechnic Institute and a Ph.D. in Ecology from the University of California, Davis.

Donald O. Thompson is Anston Marston Distinguished Professor Emeritus of Engineering (Department of Aerospace Engineering and Engineering Mechanics) at Iowa State University, where he performed research, taught new nondestructive evaluation systems, and serves as Scientific Advisor to the Director of the Institute of Physical Research and Technology at Iowa State University. He is also the Founding Director, Center for Nondestructive Evaluation, Iowa State University. Donald Thompson performed studies of radiation damage in solids and hydrogen embrittlement in metals using elastic and anelastic techniques. In both these topics, primary emphasis was placed on the development of understanding of dislocation interactions with point defects and with hydrogen. More recently, he has been extensively involved in nondestructive evaluation research. His interests include the development of new ultrasonic instrumentation for quantitative nondestructive evaluation. This work takes advantage of many theoretical advances that have been made in elastic inverse scattering, and provides ways to determine a flaw's size; shape; orientation; and composition. Previously, he was Principal Scientist and Program Director, Ames Laboratory, Applied Nondestructive Evaluation Programs; Director, Center for Nondestructive Evaluation; and Acting Director, FAA/Center for Aviation Systems Reliability. He also worked at the Rockwell International Science Center, where he held a number of positions, including: Director, Structural Materials Department; Manager, Technical Staff; Group Leader, Structural Materials; and Corporate Panelist for Nondestructive Evaluation and Structures. Prior to that, he worked at Oak Ridge National Laboratory in the Solid State Physics Division where he served as Group Leader, Elastic and Anelastic Effects and member of the Radiation Effects Group. Donald Thompson is a member of the National Academy of Engineering, a Fellow of the American Physical Society, a Life Fellow of the Institute of Electrical and Electronics Engineers, and a Distinguished Fellow of the Rockwell International Science Center. He is also an honorary Fellow of the Indian Society of Nondestructive Testing; the American Society for Nondestructive Testing; and a Foreign Associate of the Indian National Academy of Engineering. He is Senior Editor of *Proceedings of Review of Progress in QNDE*; and past Guest Editor of *Sonics and Ultrasonics*, and the First Research Supplement of *Materials Evaluation*. Donald Thompson has published over 180 papers, including those in peer-reviewed journals, and several books. He has received eight patents. Donald Thompson received his B.A., M.S., and Ph.D. degrees in Physics from the University of Iowa, Iowa City, IA.

Cheryl A. Trottier is currently Chief of the Radiation Protection, Environmental Risk, and Waste Management Branch of the Office of Nuclear Regulatory Research at the U.S. Nuclear Regulatory Commission. In that capacity, she is responsible for the management of research programs and the development of technical bases to support rulemaking. This includes the development of models for assessing the maximum doses likely to be received from lands and materials cleared from regulatory control; evaluating hydrologic model and parameter uncertainty; the development of realistic parameters for assessing sorption processes in geochemical models; and refining evaluations of radionuclide transport mechanism in the environment. In her 25 years of experience in the field of radiation protection, she has been involved in the management of environmental radiation protection monitoring programs and laboratory measurements, and the emergency preparedness coordination for an electric utility. She was also involved in the areas of materials use regulation oversight; development of regulations; and the development of guidance for use of byproduct and special nuclear materials. This included finalization of regulations and the development of regulations to certify the gaseous diffusion plants. Cheryl Trottier serves as one of the U.S. representatives to the Nuclear Energy Agency Committee on Radiation Protection and Public Health. She is a member of the American Nuclear Society, and serves as a member of the Committee on Site Clean-Up Restoration Standards. She received her B.A. degree in Biology from Rutgers University.

Wade O. Troxell is currently the director of the Robotic and Autonomous Machines (RAM) Laboratory at Colorado State University (CSU), where he is also an Associate Professor and Associate Department Head of Mechanical Engineering. At CSU, Professor Troxell investigates organizing principles underlying biological and robotic systems; synthesis of intelligent systems; and applications of intelligent control of distributed systems. His research interests consist of product realization processes; design support systems; and behavior-based robots (task-structured approach to building robust and reliable autonomous intelligent systems). His research interests also include robot programming and control (high-level formalisms; complexity measures; and verification). His professional experience is extensive and includes his positions as: Executive Director for the Colorado Manufacturing Extension Center; the interim dean of the Technical Education Division of Pueblo Community College; and Executive Director of the U.S. National Institute of Standards and Technology (NIST)/ Mid-America Manufacturing Technology Center, Colorado Regional

Office. At CSU, his positions have also included: Director of the Manufacturing Excellence Center; and Director of the Manufacturing and Robotic Systems Laboratory, Mechanical Engineering. He was also the President/Chief Operating Officer and Founder of Sixth Dimension, Inc. Additionally, he has served as: Robotic Consultant to the Public Service Company of Colorado, Nuclear Engineering Division (Fort St. Vrain Station on the controller retrofit of the fuel handling robot); NATO Postdoctoral Fellow at the University of Edinburgh for the Department of Artificial Intelligence; a Consultant specializing in product design and process automation; and a Mechanical Engineer for the Eastman Kodak Company. He has provided services as an expert witness for trade secrets, patent infringements, and product liability. He is currently Senior Vice President Elect of the American Society of Mechanical Engineers (ASME) Council of Member Affairs. He serves on the ASME Inter-Council Committee for Federal R&D; and the ASME/NIST Interaction Committee. He is on the Board of Advisors for the Northern Colorado Idea Laboratory. He is the author or coauthor of over 50 refereed publications, technical reports, and conference proceedings. Wade Troxell received a B.S. degree in Engineering Science; an M.S. degree and a Ph.D. in Mechanical Engineering, all from CSU.

Charles O. Velzy is a consultant in the field of waste treatment and disposal. Previously, he held increasingly responsible positions with the environmental consulting engineering firm, Charles R. Velzy Associates, Inc., becoming President in 1976. In 1987, when Velzy Associates merged with Roy F. Weston, Inc., Charles Velzy became Vice President of Weston, a position which he held until retiring in 1992. He has over 35 years of experience as an environmental engineering consultant specializing in: the analysis of waste management problems; design of wastewater treatment and waste disposal systems; and design of new, retrofit of existing, testing, and permitting of waste combustion facilities. He has authored or co-authored over 80 publications—primarily in the field of solid waste management. He has served on the Science Advisory Board of the U.S. Environmental Protection Agency; as President of the American Society of Mechanical Engineers (ASME); Chair of the ASME Peer Review Committee; and as Treasurer of the American

Academy of Environmental Engineers (AAEE). He has served on numerous committees of the ASME, the AAEE, the American National Standards Institute, and the American Society for Testing and Materials. He is a registered professional engineer in New York and eleven other states. Charles Velzy received B.S. degrees in Mechanical and Civil Engineering, and an M.S. in Sanitary Engineering from the University of Illinois at Urbana-Champaign.

Acronyms

ANSI	American National Standards Institute
ASME	American Society of Mechanical Engineers
CCD	charge-coupled device
CFR	Code of Federal Regulations
CMOS	complementary metal oxide semiconductor
DCEN	direct current electrode negative
DOE	U.S. Department of Energy
DOT	U.S. Department of Transportation
EM	Office of Environmental Management (DOE)
EP	Executive Panel
EMATs	Electromagnetic Acoustic Transducers
ET	Eddy Current Testing
FRR	foreign research reactors
GD&T	geometric dimensioning and tolerances
GTAW	gas tungsten arc welding
HEPA	high efficiency particulate air
HEU	highly enriched uranium
HLW	high-level radioactive waste
ICD	Interface Control Document
IGSCC	inter-granular stress corrosion cracking
INEEL	Idaho National Engineering and Environmental Laboratory
ISFSI	independent spent fuel storage installation
LMITCO	Lockheed Martin Idaho Technologies Company
MNIP	Maximum Normal In-plant Handling Pressure
MNOP	Maximum Normal Operating Pressure
MRS	monitored retrieval storage
NDE	nondestructive evaluation
NIC	Network Interface Card
NSNFP	National Spent Nuclear Fuel Program
NWPA	Nuclear Waste Policy Act of 1982
NWPAA	Nuclear Waste Policy Amendments Act of 1987
OCRWM	Office of Civilian Radioactive Waste Management
OD	outer diameter
PRCEE	Peer Review Committee for Energy and Environment
PT	project team
QA	quality assurance
RP	Review Panel

RSI	Institute for Regulatory Science
RT	radiographic test
SHIC	standard high integrity can
SNF	spent nuclear fuel
SRS	Savannah River Site
USEPA	United States Environmental Protection Agency
USNRC	United States Nuclear Regulatory Commission
UT	ultrasonic testing
VAC	volts alternating current

www.ingramcontent.com/pod-product-compliance
Lightning Source LLC
Chambersburg PA
CBHW081533220326
41598CB00036B/6418